International Explorations in Outdoor and Environmental Education

Volume 1

This series focuses on contemporary trends and issues in outdoor and environmental education, two key fields that are strongly associated with education for sustainability and its associated environmental, social and economic dimensions. It has an international focus to encourage dialogue across cultures and perspectives. The scope of the series includes formal, nonformal and informal education and the need for different approaches to educational policy and action in the twenty first century.

Research is a particular focus of the volumes, reflecting a diversity of approaches to outdoor and environmental education research and their underlying epistemological and ontological positions through leading edge scholarship.

The scope is both global and local, with various volumes exploring the issues arising in different cultural, geographical and political contexts. As such, the series aims to counter the predominantly "white" Western character of current research in both fields and enable cross-cultural and transnational comparisons of educational policy, practice, project development and research.

The purpose of the series is to give voice to leading researchers (and emerging leaders) in these fields from different cultural contexts to stimulate discussion and further research and scholarship to advance the fields through influencing policy and practices in educational settings. The volumes in the series are directed at active and potential researchers and policy makers in the fields.

More information about this series at http://www.springer.com/series/11799

Jane Edwards

Socially-critical Environmental Education in Primary Classrooms

The Dance of Structure and Agency

 Springer

Jane Edwards
RMIT University
Melbourne, VIC, Australia

ISSN 2214-4218   ISSN 2214-4226 (electronic)
International Explorations in Outdoor and Environmental Education
ISBN 978-3-319-34720-2   ISBN 978-3-319-02147-8 (eBook)
DOI 10.1007/978-3-319-02147-8

Printed on acid-free paper

Springer International Publishing AG Switzerland is part of Springer Science+Business Media
(www.springer.com)

This book is dedicated to the memory of Eric Bottomley whose contribution to a sustainable future resides in all the students who have been, and will be, touched by his work.

Foreword

This book, the first in our series, brings the ideas of Anthony Giddens' theory of structuration into the field of outdoor and environmental educational research, with the aim of assisting educational researchers, educators and educational institutions to bridge the gaps between current educational practices and the rhetoric and best practices of education for sustainable development.

Most environmental educators would agree that the implementation of environmental education programmes in many countries has often had less than ideal results, even when well supported and funded by education departments or external agencies. Many explanations have been offered for both the rare successes and many failed attempts at implementing environmental education and education for sustainable development programmes. In this book, Jane Edwards provides a well-theorized and organized argument that sheds light on the rhetoric-reality gaps in the implementation of such programmes.

Jane draws on research, guided by Giddens (1976, 1979, 1984, 1991a, b), to:

- Highlight the potential of Giddens' theory of structuration, an approach not yet well established in the field of educational research, to provide new and valuable insights into the factors and forces that shape educational practices
- Provide practical guidance to the use of this new approach to educational research
- Provide a new perspective on the nature of the gap between the rhetoric, and the reality, of education for sustainability in Australian primary schools
- Provide new perspectives on issues facing educational institutions aiming to introduce effective education for sustainable development, particularly in relation to the need for teachers to adopt new pedagogical practices

The effectiveness of educational programmes that contribute to the United Nations Decade for Education for Sustainable Development 2005–2014 (UNESCO 2005) and beyond depends on the ability of schools and teachers to embrace changes that reduce the gap between the reality of classroom practices and the rhetoric of education *for* the environment. This book responds to the need to better understand the factors that contribute to the rhetoric-reality gaps in order to inform processes for effective, and sustainable, educational change.

In this book Jane explores the experiences of teachers, from Australian primary schools, faced with the challenge of implementing a sustainable school programme. This programme positioned young students as active participants in the social processes from which environmentally sustainable practices are developed. This required teachers to adopt a socially critical pedagogy—a pedagogy that represents the antithesis of the well-established transmissive approaches they had previously enacted in environmental education.

The insights gained from the teachers' experiences provide a conceptual understanding of the classroom nature of educational rhetoric-reality gaps. These conceptual insights were derived interpretively from Anthony Giddens' theory of structuration and the notion of the duality of structure and agency—an approach to research only partially established in the field of school education writ large—and with little (if any) influence on outdoor and environmental education. Although each teacher justified their classroom practice in terms of specific aspects of structure and/or agency, it was important to explore relationships between such aspects, following Giddens. The use of hypothetical scenarios during interviews provided a neutral space for teachers to consider both personal and external factors that influenced their pedagogical choices and highlighted the role of various unstated, unconscious and nonconscious ideas in those choices. This provided unique perspectives into the teacher-mediated manner in which certain elements of structure and agency interrelate to enable and constrain classroom practices.

These interrelationships of agency and structure in the classroom provide valuable starting points for future research to identify ways to reduce educational rhetoric-reality gaps and inform processes for effective educational change. Most significantly, this book demonstrates that Giddens' theory of structuration is a valuable framework that can contribute to the development of new understandings in the field of educational research.

We are delighted that Jane's work is the first book in our series as we have had a long association of working together. Annette first encountered Jane as a graduate teacher education student when we were at Deakin University. As a result of her geological research background, she impressed Annette with her writing and research abilities, and Annette suggested that she do her PhD. Jane was an active member of the Gough Girls, our combined doctoral student seminar group, and then she followed Annette to RMIT where she also became a lecturer in science education for a time, before moving to the Australian Council for Educational Research.

RMIT University, Melbourne, VIC, Australia   Annette Gough
La Trobe University, Melbourne, VIC, Australia   Noel Gough

References

Giddens, A. (1976). *New rules of sociological method*. New York: Hutchinson.

Giddens, A. (1979). *Central problems in social theory: Action, structure and contradiction in social analysis*. London: MacMillan.

Giddens, A. (1984). *The constitution of society*. Cambridge: Polity Press.

Giddens, A. (1991a). *Modernity and self-identity. Self and society in the late modern age*. Cambridge: Polity Press.

Giddens, A. (1991b). Structuration theory: Past, present and future. In C. G. A. Bryant, & D. Jary (Eds.), *Giddens' theory of structuration: A critical appreciation* (pp. 201–221). London: Routledge.

UNESCO. (2005). *United Nations Decade of Education for Sustainable Development 2005–2014. International implementation scheme*. United Nations Educational, Scientific and Cultural Organisation. http://unesdoc.unesco.org/images/0014/001486/148654e.pdf. Accessed 1 Oct 2014.

Preface

How to best prepare children to participate in, contribute to, and confidently embrace the changing opportunities of our rapidly evolving global community is a challenge faced by every teacher in every school. Despite attempts to implement educational programs that aim to assist teachers to successfully respond to such a challenge, educational leaders are often frustrated by the apparent inability or reluctance of many teachers to adopt new approaches. This book will assist those who contribute to educational policy and practice, whether through school leadership, classroom teaching, curriculum design or educational research, to more effectively facilitate educational improvement. New insights into how to interpret pedagogical practice, particularly in relation to guiding research into teachers' pedagogical choices and informing the facilitation of pedagogical change in the classroom, will improve the effectiveness of programs that embrace the educational practices that best prepare today's students for their future roles in society.

Today, I envisage humanity as sitting on the edge of a precipice and faced with making decisions that will undoubtedly influence the viability of life on Earth. Failure to acknowledge and act upon the evidence that the environmental consequences of current human–environment relationships are unsustainable, and indeed detrimental to human life, will undoubtedly lead to an irreversible plummet: a rapid decline in life quality caused by a cascade of global environmental changes unprecedented in human history. On the other hand, the decision to embrace the notion of sustainability and act now to transform human–environment relationships may enable humanity to take one step back from the edge of this precipice. Irrespective of decisions made today, the enormity of the survival challenges yet to be faced by humanity is seemingly overwhelming. Despite this, I strongly believe that, with a little ingenuity, every individual can influence humanity's journey into the future: a journey that will be shaped, in part, by the way in which today's educators prepare students for their future decision-making roles. The information and understandings expressed in this book represent part of my contribution to that journey.

The classroom practices of my school education, and therefore the social values and educational goals embedded within them, were not dissimilar to those experienced by my parents and indeed strikingly similar to many of those that I observe in today's classrooms. This seems quite amazing. In light of increasingly rapid changes in human–environment relationships, how can those values and goals established by previous generations prepare today's students to deal with humanity's future challenges? I believe that educational institutions have a vital role to play in assisting any one generation to not just understand and respect the wisdom of previous generations but to build on that wisdom in ways that empower individuals to actively participate in improving their society. It is important that all who contribute to educational policy and practice, whether through regulation, research, or teaching, continuously question the role of education in society, their role in shaping that education, and whether or not their favoured educational practices meet both the current and future needs of students. In the first chapter I discuss the questions that I believe must be asked if education is to empower the next generation to contribute to the transformation of the human–environment relationships that, if left unchecked, will undoubtedly cause social and environmental devastation. Of course, asking the right questions is just the beginning of any journey—transformation requires finding answers and taking appropriate action. The action I focus on here is classroom practice, or pedagogy, as irrespective of the intentions of educational policy developers, or the recommendations of educational researchers, it is only through teachers' practices that educational change actually occurs. The prevalence of educational rhetoric–reality gaps in today's classrooms, that is, the difference between the values and goals embedded in a teacher's practice compared with the values and goals embedded in practices advocated by policy or programs, suggests that transforming educational practices is not a simple process.

I believe that the first part of effectively questioning educational practice, whether you are a policy developer or a classroom teacher, is to have an understanding of the social and educational history of the practices to be questioned. It is only through questioning past practices that it is possible to identify ways in which to successfully embrace the values and goals embedded in new practices. In Chap. 2, I outline the historical, social, and educational context for the development of the Sustainable Schools Program, an exemplar of education for sustainable development in Australia, and the socially-critical pedagogy that the program advocates. However, successfully implementing educational change requires much more understanding that can be gleaned from the documents that describe how and why a new practice is appropriate—it requires an understanding of what actually influences a teacher's actions in the classroom. And what influences a teacher's actions cannot be simply understood in terms of a particular educational regulation or specific individual motivation; after all, if it were that simple, there would be many fewer educational rhetoric–reality gaps.

As I reflect on my journey towards writing this book, my gaze turns to the bookshelf crammed with an assortment of well-thumbed dog-eared books and filing cabinet drawers that overflow with handwritten notebooks and photocopied documents pockmarked with highlighted passages and scores of fading post-it notes with quickly scribbled ideas. This scene is testament to my early attempts to understand each and every aspect of Anthony Giddens' theory of structuration and the notion of the duality of structure and agency, as I grappled with my decision to develop an understanding of the development of rhetoric–reality gaps through a theory for which there was no established epistemological procedures, few examples of its use in the field of educational research, and a multitude of reports warning that Giddens' ideas were notoriously difficult to read and comprehend. In Chap. 3, I present my interpretation of the ideals of Giddens' theory of structuration and highlight how these relate to, and indeed provide new perspectives of, educational practice. I encourage everyone who in any way contributes to educational policy or practice to explore Giddens' ideas with a view to better understand the social interactions that constitute education and therefore, in turn, better understand how to more effectively facilitate educational change. In Chap. 4, I outline the ways in which I adapted Giddens' theory of structuration to the field of educational research. I strongly believe that research informed by the ideals of structuration has the potential to expose perspectives and ideas about the interplay of structure and agency within an educational setting that may otherwise have been masked by more traditional approaches. I strongly encourage educational researchers to explore ways in which to employ the theory of structuration in their investigations. I believe that those of us who contribute to the field of educational research must find ways in which to embrace change in order to find more effective ways in which to inform and facilitate the appropriate transformation of educational practice into the future.

The need for educational research to provide a better understanding of educational practice and the demands of educational change is demonstrated by the teachers' stories that I present in Chap. 5. These stories record the experiences and perspectives of the teachers implementing the Sustainable Schools Program and are instructional as a snapshot of the range of classroom practices that can result from a single educational directive. The presence of significant educational rhetoric–reality gaps in education for sustainable development in Australia is indisputable.

My use of Giddens' theory of structuration to develop an understanding of this educational issue, particularly in terms of identifying the sociocultural and hermeneutic factors that most influence a teacher's classroom practice, is incorporated into the discussions of Chaps. 6, 7, 8 and 9. These chapters are obviously instructive for those educational researchers looking for examples of how to relate Giddens' ideas to real-life situations. However, I also encourage other educational practitioners to understand pedagogy from the perspective of Giddens' theory of structuration, particularly in terms of the ways in which relationships between sociocultural and hermeneutic factors influence teachers' classroom practices. If you are a teacher, these chapters may increase your awareness of why you teach the way that you do or highlight why your colleagues prefer different approaches. If you are a principal, these chapters will highlight why some of your attempts to alter the practices in your school have, or have not, been successful and might assist you to devise new and more effective strategies for implementing change. Similarly, if you are involved in the development of educational policy or curriculum design, and particularly if you wish to influence pedagogy, these chapters may highlight the factors of educational practice that *must* be addressed if your educational plans aim to facilitate long-lasting change.

Facilitating effective and long-lasting educational change is an enormous under-taking, for which there is no single or simple procedure. This book provides just a snapshot of teachers' pedagogical practices, the development of educational rheto-ric–reality gaps, and the application of an underutilized theory to educational research during just one process of educational change. The insights offered here provide no definitive solution for the reduction of educational rhetoric–reality gaps, but highlight important ideas to contribute to building a better understanding of how to more effectively facilitate educational, and therefore societal, transformation.

RMIT University, Melbourne, VIC, Australia   Jane Edwards

Acknowledgement

I owe my deepest gratitude to Professor Annette Gough. Annette's contribution to environmental education inspired me to begin the doctoral research that is reported in this book. Annette's encouragement, guidance and friendship not only kept my research on track but also ensured that I learned much along the way. I am heartily thankful to Professor John Fien, whose advice shaped my approach to the research that is reported in this book. John's enthusiasm and willingness to share his ideas encouraged me to delve into methodological murky waters, a journey for which I am now thankful to have taken.

Many thanks to the school principals who granted me access to their schools and permission to approach staff at a very busy time of the year. I am most indebted to the teachers who invited me into their classrooms and so openly shared with me their ideas, fears, and hopes. Their passion for teaching and commitment to their students was inspiring.

Many people have offered valuable advice and unselfishly given some of their time to assist me with this work and, most importantly, forced me to think a little "outside the box". In particular, I wish to acknowledge these contributions from Professor Noel Gough, Professor Russell Tytler, Glenn Davidson, Eric Bottomley, and Pat Armstrong. Many thanks also to Larraine Larri for assisting to locate certain references and permission to publish the tables in Chap. 2.

I offer my warmest thanks to Ross, Robyn, Jenni, and Helen and to the many friends and colleagues who have supported me through this research and book writing journey.

Contents

Abbreviations

AAEE	Australian Association for Environmental Education
AARE	Australian Association for Research in Education
ACER	Australian Council for Educational Research
AuSSI	Australian Sustainable Schools Initiative
CDC	Australian Curriculum Development Centre
CERES	Centre for Education and Research in Environmental Strategies
CRT	Casual relief teacher
DEH	Department of the Environment and Heritage
DESD	Decade of Education *for* Sustainable Development
DEST	Department of Education, Science and Training
DET	Department of Education and Training
DHAE	Department of Home Affairs and Environment
DSE	Department of Sustainability and Environment
EE	Environmental education
EPA	Environmental Protection Authority
ESD	Education for sustainable development
EVNP	East Valley Nature Park
ICT	Information and communication technology
IEEP	International Environmental Education Programme
IT	Information technology
IUCN	International Union for the Conservation of Nature
JPOI	Johannesburg Plan of Implementation
LOTE	Language other than English
MCEETYA	Ministerial Council on Education, Employment, Training and Youth Affairs
MEAB	Millennium Ecosystem Assessment Board
OECD	Organisation for Economic Co-operation and Development
OHS	Occupational Health and Safety
PD	Professional development

PE	Physical education
SOSE	Studies of society and the environment
SSP	Sustainable Schools Program
UNCED	United Nations Conference on Environment and Development
UNEP	United Nations Environment Programme
UNESCO	United Nations Educational, Scientific and Cultural Organization
VELS	Victorian Essential Learning Standards
WCED	World Commission on Environment and Development
WSSD	World Summit on Sustainable Development
WWF	World Wildlife Fund for Nature

List of Figures

List of Tables

Chapter 1
The Rhetoric–Reality Gap

The understanding that it is not possible to sustain current human–environment relationships, and that the social and environmental consequences of unmitigated use of natural resources and exponential population growth are catastrophic, has led to global calls to transform the way that human societies operate. Any journey of transformation begins with the willingness and ability to question the philosophy upon which current practices are founded. This means that if a society is to transform the well-established human–environment relationships that define, and are defined by, the cultural values of that society, significant questioning must take place: the questioning of the values that shape the way in which that society operates; the questioning of the role of educational institutions and how these support either the continuance, or the transformation, of that society's predominant cultural values; the questioning of the practices of educators and the role they play in shaping and empowering that society's future decision makers; and the questioning of how an understanding of these issues is best developed in order to most effectively inform the process of transformation.

1.1 The Questioning of Society

Archaeological evidence and cultural stories from the past 1000 years of human history contain enumerable examples of the breakdown of human–environment relationships due to an insatiable need and/or desire for natural resources (e.g. Butti and Perlin 1980; Diamond 2005; Grove 1995; Miller et al. 2005; Lee and Williams 2001). The seemingly unmitigated use of natural resources that characterise the predominant human–environment relationships of the industrialised societies today strongly reflects one of the most radical changes to the manner in which humans perceive and interact with their environment in modern human history—the replacement of mysticism with science as the primary vehicle for understanding the world

© Springer International Publishing Switzerland 2016

J. Edwards, *Socially-critical Environmental Education in Primary Classrooms*,
International Explorations in Outdoor and Environmental Education 1,
DOI 10.1007/978-3-319-02147-8_1

by western civilisations[1] (Gribbin 2002). The widespread acceptance of rational thought and a scientific worldview enabled the rapid development of industrial and technological achievements, and an insatiable appetite for energy. Today, progress is often measured by industrial and technological advancement rather than human development, and life quality is often aligned to the capitalist political values that contribute to an individual's economic capacity to consume the products of such advancement (O'Sullivan 1999; Spring 2004; Swain 2005). Today's industrialised societies enjoy the benefits of valuable advancements, such as those in the field of medical science, alongside an unprecedented level of per capita consumption that is well in excess of basic survival needs. And not only are individuals consuming more, there are increasingly more consuming individuals.

During the last half a century, more people have been born than were born during the previous four million years. During the next half a century, it is likely that the Earth's population will treble in size to over nine billion (Kunzig 2011). During the last half a century, the ecological footprint[2] of humanity increased from fifty percent of the Earth's available capacity to a staggering thirty percent more than the Earth's capacity. During the next half a century, it is estimated that the resources of more than two Earths will be required to support the human population (Wackernagel and Rees 1998; WWF 2012). The effects of capitalist-based social values, in conjunction with such unrestrained population growth, has facilitated unmitigated natural resource use that is evidenced today by increasing instability within and between Earth's ecosystems (MEAB 2005). Since the early 1970s there has been a growing social awareness that increased rates of population growth and unmitigated resource use are not sustainable, as these have facilitated a significant decline in the capacity of Earth systems to continue to support human needs (Suzuki 2003).

Educational institutions traditionally adopt practices that reflect and support the predominant cultural values of the society in which they operate. The capitalist values of the industrialised societies are reflected in educational practices that have facilitated a gradual but significant disenfranchisement of individuals from the natural components of the world. O'Sullivan (1999) notes that at any time when the overriding or dominant cultural basis of a society is "at its zenith" there is no sense that change is needed, and correspondingly there is little questioning of the educational foundations of society, as such questioning requires acknowledgement that what has been, and indeed what is, is perhaps no longer appropriate (O'Sullivan 1999, p. 4). Today however, there is growing social awareness that the dominant human-centred approach to life on Earth, fuelled for so long by the economic prosperity brought by rapid and extensive industrial and technological development, is not sustainable and has definable limits that are fast approaching. The scale of the detrimental effects of many human–environment relationships is well demonstrated

---

[1] The term 'western science' refers here to the Anglo-Saxon led developments beginning around the mid-1500s, and disregards earlier, albeit significant, developments of the ancient Greek, Chinese and Islamic scientists.

[2] The term ecological footprint refers here to the land area required to supply the resources used by humanity: the lifecycle of all products and services.

by global warming: "today we are hearing and seeing dire warnings of the worst potential catastrophe in the history of human civilization: a global climate crisis that is deepening and rapidly becoming more dangerous than anything we have ever faced" (Gore 2006, p. 10). This means that the purposes and practices of education must be questioned.

1.2   The Questioning of Education

As humanity celebrated the start of the twenty-first century, the United Nations Secretary-General Kofi Annan called for an assessment of "the consequences of ecosystem change for human well-being and the scientific basis for actions needed to enhance the conservation and sustainable use of those systems and their contribution to human well-being" (MEAB 2005, p. 1). Five years later, the Millennium Ecosystem Assessment Board concluded that:

> it lies within the power of human societies to ease the strains we are putting on the natural services of the planet, while continuing to use them to bring better living standards to all. Achieving this, however, will require radical changes in the way nature is treated at every level of decision-making. Resilience and abundance can no longer be confused with indestructibility and infinite supply. The warning signs are there for all of us to see. The future lies in our hands (MEAB 2005, p. 23).

And so began the Decade of Education *for* Sustainable Development (DESD), 2005–2014, which embraced the notion that responding to the problems caused by current human–environment relationships requires widespread and long lasting social transformation, and that this is best achieved through appropriate education of society's future decision makers (WCED 1987; WSSD 2002).

O'Sullivan (1999) considered the notion of Education *for* Sustainable Development (ESD) to represent a "transformative moment" (p. 5) in that it questioned not only the value and sustainability of the well-established human–environment relationships upon which today's industrialised societies are founded, but also the educational practices that supported and ensured continuity of those relationships. Such educational practices reflect a 450-year history of development alongside the industrial and technological advancements that have been driven by the findings of scientific methodologies and the notion of positivist inquiry.

Altering the human–environment relationships of society however, requires acknowledgement that educational systems reflect what is essentially an "outdated worldview" or inadequate "perception of reality" (Capra 1997, p. 4) because there are "fundamental differences in goals, content, and methods between the educational interests of the nation-state, neoliberal ideas, and human rights and environmental education agendas" (Spring 2004, p. 164). The positivist focus on knowledge that forms the basis of the vocational/neo-classical pedagogy of traditional science education fails to embrace the future-oriented and socially transformative goals of ESD—goals best achieved through a socially-critical pedagogy (e.g. Fien 2001; 1993; Gough 1997; Scott and Gough 2003, 2004; Tilbury et al. 2002). In other words, educational practices must change in order to facilitate the establishment of more sustainable human–environment relationships (WCED 1987; WSSD 2002). The challenge for educators is to identify how the existing philosophies of educational institutions might be modified to better support current and future social, and therefore educational, needs. The implementation of education that aspires to transform society requires finding ways to bridge the gaps between past and present wisdoms, and future needs. Despite over forty years of calls for practices in schools to more effectively support the socially transformative goals of environmental education, and most recently of ESD, traditional vocational/neo-classical pedagogies remain predominant (McKeown 2002). This supports the notion that the transformation of educational practices is somewhat problematic (e.g. Andrews 1996; Fullan 2003, 2007; Hargreaves 1997; Sarason 1990). As noted by Donnison (2004), "teachers and educational institutions are resistant to change" (p. 26), in part, because "the way that teachers are trained, the way that schools are organised, the way that the educational hierarchy operates…results in a system that is more likely to retain the *status quo* than to change" (Fullan 2003, p. 3, original italics). The ideals of ESD represent a challenge to educators to develop new ways to think about their role in education. The inability to readily embrace such change and to re-imagine the purposes and practices of education is illustrated by the development of educational rhetoric–reality gaps.

1.3   The Questioning of Educators

Rhetoric–reality gaps, named by Stevenson (1987; 2007) to describe what had been a long-recognised phenomenon, have been well-documented by environmental educators as the differences between the rhetoric of educational theory and the reality of a teacher's pedagogy (e.g. Fien 2001; Tilbury et al. 2004). Many educational rhetoric–reality gaps have developed in response to calls to depart from a knowledge-based vocational/neo-classical pedagogy in order to accommodate the goals of environmental education through a socially-critical pedagogy (e.g. Bishop and Russell 1985; Fien 2001; McKeown 2002; Robertson and Krugly-Smolska 1997; Stapp and Stapp 1983; Stevenson 2007). In Victoria, the Australian Sustainable Schools Initiative (AuSSI) aimed to guide educational transformation towards effective ESD through the implementation of the Sustainable Schools Program (SSP). Achievement of the future-oriented and socially-transformative goals of this program was contingent upon the use of a socially-critical pedagogy. In light of the need for teachers to employ a socially-critical pedagogy, the presence of rhetoric–reality gaps in the implementation of SSP was not unexpected.

If education is to contribute to the establishment of sustainable human–environment relationships, it is important to identify ways in which to reduce the development of rhetoric–reality gaps in order to assist teachers to more effectively transform the rhetoric of ESD, as represented in programs such as SSP, into the reality of their classroom practices. This requires developing a holistic understanding of educational rhetoric–reality gaps, and their relationship to the manner in which teachers approach educational change, and in particular, pedagogical change. In other words, informing ways in which to begin to improve the implementation of effective ESD requires finding answers to two questions, namely, (i) what rhetoric–reality gaps are significant for ESD? and (ii) how can these rhetoric–reality gaps be reduced? The presence of rhetoric–reality gaps in the implementation of SSP and a socially-critical pedagogy provided opportunities to answer these questions through an investigation of teachers' actions to implement educational change. The methodological issues, teachers' stories and understandings of the development of rhetoric–reality gaps related to this investigation are reported in this book. This investigation encompassed the following three aspects of the educational change process required to implement SSP:

- Rhetoric—the teachers' understandings of the environmental ideologies embraced by the goals of SSP and the educational ideologies represented by the practices of a socially-critical pedagogy;
- Reality—the manner in which the teachers incorporated SSP into their classroom practices; and
- Teachers' experiences—the practicalities of implementing new practices within the well-established routines of an educational institution.

1.4   The Questioning of Understanding

When undertaking an investigation into a well-documented phenomenon that has persisted over a significant period of time it is important to consider how to gain new perspectives and ideas that could contribute to moving forward. This requires researchers to engage with new and different methodologies in the search for previously unseen perspectives that may lead to new understandings. In light of this, the use of Anthony Giddens' theory of structuration for example, an approach yet to be established within the field of educational research, offers the potential to reveal new insights into the issues faced by teachers when asked to change their well-established educational practices in order to implement new programs, such as SSP and the associated socially-critical pedagogy. Such new insights have the potential to inform new ways of thinking about the practice of education, to identify ways in which to reduce the development of educational rhetoric gaps to improve ESD, and to demonstrate the value of applying new methodological approaches to educational research.

References

Andrews, C. (1996). *Teachers' work: An analysis of teachers' work in a context of change*. PhD, Griffith University, Queensland, Australia.

Bishop, J., & Russell, G. (1985). Study tours in Australia and the USA. *Review of Environmental Education Developments, 13*(2), 14–15.

Butti, K., & Perlin, J. (1980). *A golden thread: 2500 years of solar architecture and technology*. New York: Van Nostrand Reinhold.

Capra, F. (1997). *The web of life: A new synthesis of mind and matter*. London: Flamingo.

Diamond, J. (2005). *Collapse. How societies choose to fail or survive*. Melbourne: Penguin Group.

Donnison, S. (2004). The 'digital generation', technology, and educational change: An uncommon vision. In B. Bartlett, F. Bryer, & D. Roebuck (Eds.), *Educating: Weaving research into practice* (Vol. 2, pp. 22–31). Brisbane: School of Cognition, Language and Special Education, Griffith University.

Fien, J. (1993). *Education for the environment: Critical curriculum theorising and environmental education*. Geelong: Deakin University Press.

Fien, J. (2001). *Education for sustainability: Reorientating Australian schools for a sustainable future* (Tela series). Fitzroy: Australian Conservation Foundation. http://www.acfonline.org.au/sites/default/files/resources/tela08_education_%20for_sustainability.pdf. Accessed 30 Sept 2014.

Fullan, M. (2003). *Change forces: Probing the depths of educational reform* (3rd ed.). London: Falmer.

Fullan, M. (2007). *The new meaning of educational change* (4th ed.). New York: Teachers College Press.

Gore, A. (2006). *An inconvenient truth. The planetary emergency of global warming and what we can do about it*. London: Bloomsbury.

Gough, A. (1997). *Education and the environment: Policy, trends and the problems of marginalisation*. Melbourne: Australian Council for Educational Research.

Gribbin, J. (2002). *The scientists: A history of science told through the lives of its greatest inventors*. New York: Random House.

Grove, R. H. (1995). *Green imperialism: Colonial expansion, tropic island Edens and the origins of environmentalism, 1600–1800*. Cambridge: Cambridge University Press.

Hargreaves, A. (1997). Cultures of teachers and educational change. In M. Fullan (Ed.), *The challenge of school change: A collection of articles* (pp. 47–69). Arlington Heights: IRI/Skylight Training and Publishing.

Kunzig, R. (2011). Population 7 billion. *National Geographic, 219*(1), 32–69.

Lee, J. C.-K., & Williams, M. (2001). Researching environmental education in the school curriculum: An introduction for students and teacher researchers. *International Research in Geographical and Environmental Education, 10*(3), 218–244.

McKeown, R. (2002). *Education for sustainable development toolkit*, version 2, July 2002. Energy, Environment and Resources Center, University of Tennessee, USA. http://www.esdtoolkit.org. Accessed 30 Sept 2014.

MEAB. (2005). *Living beyond our means. Natural assets and human wellbeing*. A report of the Millennium Ecosystem Assessment Board, United Nations Environment Programme. http://www.maweb.org/documents/document.429.aspx.pdf. Accessed 30 Sept 2014.

Miller, G. H., Fogel, M. L., Magee, J. W., Gagan, M. K., Clarke, S. J., & Johnson, B. J. (2005). Ecosystem collapse in Pleistocene Australia and a human role in megafaunal extinction. *Science, 309*, 287–290.

O'Sullivan, E. (1999). *Transformative learning: Educational vision for the 21st century*. Toronto: University of Toronto Press.

Robertson, C. L., & Krugly-Smolska, E. (1997). Gaps between advocated practices and teaching realities in environmental education. *Environmental Education Research, 3*(3), 311–326.

Sarason, S. B. (1990). *The predictable failure of educational reform: Can we change course before it's too late?* San Francisco: Jossey-Bass.

Scott, W., & Gough, S. (2003). *Sustainable development and learning: Framing the issues.* London: Routledge.

Scott, W., & Gough, S. (2004). *Key issues in sustainable development and learning: A critical review.* London: Routledge.

Spring, J. (2004). *How educational ideologies are shaping global society: Intergovernmental organisations, NGOs, and the decline of the nation-state.* Mahwah: Lawrence Erlbaum.

Stapp, W. B., & Stapp, G. L. (1983). A summary of environmental education in Australia. *Australian Association for Environmental Education Newsletter, 12,* 4–6.

Stevenson, R. B. (1987). Schooling and environmental education: Contradictions in purpose and practice. In I. Robottom (Ed.), *Environmental education: Practice and possibility* (pp. 69–82). Burwood: Deakin University Press.

Stevenson, R. B. (2007). Schooling and environmental/sustainability education: From discourses of policy and practice to discourses of professional learning. *Environmental Education Research, 13*(2), 265–285.

Suzuki, D. (2003). *A David Suzuki collection. A lifetime of ideas.* St Leonards: Allen and Unwin.

Swain, A. (2005). *Education as social action.* London: Palgrave MacMillan.

Tilbury, D., Hamu, D., & Goldstein, W. (2002). Learning for sustainable development. In *'Education' earth year report on the world summit* (pp. 9.0–9.3). Gland: International Union for the Conservation of Nature.

Tilbury, D., Coleman, V., & Garlick, D. (2004). *A national review of environmental education and its contribution to sustainability in Australia.* Canberra: Report prepared by Macquarie University for The Department of Environment and Heritage.

Wackernagel, M., & Rees, W. (1998). *Our ecological footprint: Reducing human impact on the earth.* Gabriola Island: New Society Publishers.

WCED. (1987). *Report of the World Commission on Environment and Development: Our common future.* http://www.un-documents.net/wced-ocf.htm. Accessed 30 Sept 2014.

WSSD. (2002, August 26–September 4). *The Johannesburg plan of implementation.* The World Summit on Sustainable Development, Johannesburg (South Africa). http://www.un-documents.net/jburgdec.htm. Accessed 30 Sept 2014.

WWF. (2012). *Living planet report 2012. Biodiversity, biocapacity and better choices.* World Wildlife Fund for Nature International, Gland, Switzerland. http://d2ouvy59p0dg6k.cloudfront.net/downloads/1_lpr_2012_online_full_size_single_pages_final_120516.pdf. Accessed 30 Sept 2014.

Chapter 2
Facing the Future

International calls for the immediate implementation of Education *for* Sustainable Development (ESD), as an urgent response to the global-scale environmental crises developing from current unsustainable human–environment relationships, face the paradox that educational systems are notoriously slow and difficult to alter. This chapter identifies the educational rhetoric associated with ESD by briefly outlining the 40-year journey from traditional, science-based environmental education to ESD, as it occurred in Australia in response to international recommendations. Important pedagogical responses to changes to the perceived needs and outcomes of environmental education are highlighted, with particular emphasis on the role of pedagogical practice. Effective ESD demands a socially-critical pedagogy, the goals and practices of which represent the antithesis of well-established classroom approaches into which environmental education has been traditionally slotted. Of significant concern is that the calls for educational change will simply contribute to the ever-widening gap between the reality of classroom practices and the rhetoric of education *for* the environment. The development of the Australian Sustainable Schools Initiative (AuSSI) is introduced as an exemplar of the requirement to implement ESD through a socially-critical pedagogy in Victoria. In particular, the current status of ESD is assessed in terms of the ways in which schools and teachers are implementing it, and the need to broaden educational research methods in order to better understand the issues that continue to thwart its effective implementation.

2.1 International Recommendations for Environmental Education

During the 1970s, evidence that human–environment relationships, particularly the unmitigated overuse of natural resources, were critically endangering Earth's environmental systems began to gain widespread public attention. This led to calls for

© Springer International Publishing Switzerland 2016
J. Edwards, *Socially-critical Environmental Education in Primary Classrooms*,
International Explorations in Outdoor and Environmental Education 1,
DOI 10.1007/978-3-319-02147-8_2

environmental education through which such well-established human–environment relationships would be transformed, and impending social and environmental crises averted. In 1970, the International Union for the Conservation of Nature and Natural Resources (IUCN) Nevada conference concluded that "environmental education was a science-orientated multi-disciplinary subject where most, if not all, school subjects could, and should be, incorporated" (Martin 1975, p. 21). Environmental education was viewed as a process which provided students with opportunities for:

> recognising values and clarifying concepts in order to develop skills and attitudes necessary to understand and appreciate the interrelatedness among man [sic], his culture and his bio-physical surroundings. Environmental education also entails practice in decision-making and self formulating of a code of behaviour about issues concerning environmental quality (quoted in A. Gough 1997, p. 8).

In 1972, recommendations for the establishment of the UNESCO-UNEP International Environmental Education Programme (IEEP) at the United Nations conference on the Human Environment in Stockholm more clearly positioned environmental education as a means for encouraging people to take action according to their developing 'codes of behaviour':

> Education and training on environmental problems are vital to the long-term success of environmental policies because they are the only means of mobilising an enlightened and responsible population, and of securing the manpower needed for practical action programmes (quoted in Gough 1997, p. 3).

Linke (1980) noted that by the mid-1970s, international calls for environmental education identified several critical educational outcomes directed towards developing a society's understanding of (i) human–environment relationships and human influence on environmental systems and (ii) their responsibility for ensuring quality of human life while actively contributing to environmental conservation (see also Gough 1997). The IEEP supported the development of these outcomes into more substantial policies at the International Environmental Workshop in Belgrade (in the former Yugoslavia) in 1975. Here, for the first time, a global framework (the Belgrade Charter) was provided for the most important goals of effective environmental education:

> The goal of environmental education is to develop a world population that is aware of, and concerned about, the environment and its associated problems, and which has the knowledge, skills, attitudes, motivations, and commitment to work individually and collectively toward solutions of current problems and the prevention of new ones (UNESCO 1975, p. 3).

The Belgrade Charter incorporated the growing understanding that humans needed to transform the manner in which they interacted with their environments. In particular, the charter demonstrated the understanding that environmental education must ensure that individuals are able and willing to take positive action in ways that benefit both humans and the environment. These broad statements were more fully developed during the 1977 Intergovernmental Conference on Environmental Education in Tbilisi, USSR, and presented as the Tbilisi Declaration (Gough 1997):

Environmental education, properly understood, should constitute a comprehensive lifelong education, one responsive to changes in a rapidly changing world. It should prepare the individual for life through an understanding of the major problems of the contemporary world, and the provision of skills and attributes needed to play a productive role towards improving life and protecting the environment with due regard given to ethical values (UNESCO 1978, p. 24).

This Declaration positioned environmental education as a future-oriented, global and interdisciplinary lifelong process of learning which values cooperation in the prevention and solution of environmental problems. It noted that in order to ensure individuals are able and willing to take action, environmental education must embrace four specific goals: awareness, knowledge, attitudes, skills and participation. Furthermore, these goals could only be achieved through a holistic approach encompassing economic, political, cultural-historical, ethical and aesthetic perspectives. Unlike many earlier statements, this declaration also acknowledged the importance of pedagogy in achieving environmental education goals. It indicated that learners must be assisted to develop critical thinking and problem-solving skills by becoming active participants in "planning their learning experiences...making decisions and accepting their consequences" particularly within their local environment, such that environmental education must "utilize diverse learning environments and a broad array of educational approaches to teaching, learning about and from the environment with due stress on practical activities and first-hand experience" (UNESCO 1978, p. 27). Most significantly, the Tbilisi Declaration validated the need for critical reflection of established human–environment relationships, and unquestionably acknowledged the need for significant societal transformation.

UNESCO-UNEP has reviewed the progress of the international implementation of the Tbilisi Declaration on several occasions. The 1987 conference in Moscow developed an International Strategy for Action in the Field of Environmental Education and Training for the 1990s (A. Gough 1997). The 1997 conference in Thessaloniki focused on "Education and Public Awareness" as critical for effective implementation of the Tbilisi principles. The Declaration of Thessaloniki recommended that decisions and actions of international, national and local social interactions must give "priority to education, public awareness and training for sustainability" (UNESCO 1997a, p. 3). Recommendations arising from the 2007 conference in Ahmedabad reflected the increasing understanding of "the harsh reality that not only are we exhausting and plundering the resources of the Earth at unsustainable rates, but we are on the threshold of unimaginable devastation that climate change is likely to bring" (UNESCO 2007, pp. 3–4), and that this demands urgent social transformation:

We no longer need recommendations for incremental change; we need recommendations that help alter our economic and production systems, and ways of living radically. We need an educational framework that not only [facilitates] such radical changes, but can take the lead (UNESCO 2007, p. 4).

All of these conferences reaffirmed the environmental education principles, established by the Tbilisi Declaration, which have endured as the framework for environmental education in Australia and around the world (Gough 1997; Fien

2001). However, the Ahmedabad Declaration most clearly articulated a sense of urgency for social transformation, and called for urgent changes to the purpose and practices of education: "fundamental changes in the creation, transmission and application of knowledge in all spheres and at all levels" (UNESCO 2007, p. 4).

2.2   Environmental Education in the Classroom

Environmental education began to be more widely practiced during the late 1970s–1980s but early attempts rarely addressed the full spectrum of learning outlined by the Tbilisi Declaration. In general, existing science curricula were modified to incorporate discrete ecological and conservation topics in order to educate about the natural environment. With "roots in the scientific paradigm" such environmental education remained "relatively impervious to cross-disciplinarity, and engagement with political, historical, and cultural questions" (Matthews 2011, p. 270). This science-based approach valued knowledge and awareness, rather than attitudes, skills or participation, in the belief that these alone would enable society to reduce the degradation of Earth's environmental systems (Orr 1999; Spring 2004). This reflects the belief that there is a strong relationship between awareness and knowledge, critical reflection and behaviour modification. The fact that, in general, more highly educated nations have the largest ecological footprints (WWF 2012) demonstrates that such relationships, at least in relation to environmental education, are complex and unpredictable (Kollmuss and Agyeman 2002). It has also been shown that "too much environmental knowledge (particularly relating to the various global crises) can be disempowering, without a deeper and broader learning process taking place" that enables students to respond, through action, to their developing awareness and understanding (Sterling 2003, p. 19). In other words, appropriate pedagogy is central to achieving effective environmental education: a notion addressed by Lucas (1972, 1979) in the development of his tripartite model for environmental education.

2.2.1   The Lucas Model

During the early 1970s, a review of the content and intended outcomes of environmental education practices in Australia identified three common themes:

- awareness of interrelationships between man [sic] and the environment, and the understanding of both the nature and implications of human impact on the environment;
- a concern for the quality of human life; and
- the promotion of a personal commitment to, or acceptance of responsibility for, environmental conservation (Linke 1980, pp. 27–34).

At that time, Linke (1980) identified Lucas' (1972, 1979) tripartite environmental education model as most comprehensively representing the multifaceted practices of environmental education. Lucas' model aimed to "reduce the ambiguity of the term 'environmental education'" by representing the goals of different components of environmental education which he termed education *in*, *about* and *for* the environment (Thomas 2005, p. 107). Education *about* the environment, that is, the development of "cognitive understanding" and the "development of skills necessary to obtain this understanding" (Lucas1980, p. 167) had long been well represented as science education. Education *in* the environment referred to experiential learning during which instruction occurred "outside the classroom" in the "biophysical and/or social context in which groups of people exist", while education *for* the environment was "directed to environmental preservation or improvement for particular purposes" (Lucas 1980, p. 167). Lucas argued that the process of learning was just as important as the content learned: education *in* the environment encouraged learning that engaged "all the senses, not just the intellect", whereas education *for* the environment encouraged active and contextually appropriate experiential learning (Orr 1999, p. 234).

The validity of each of the three components of Lucas' model has been extensively debated (e.g. Fien 1993; Gayford 1996; Gough 1997; Jickling and Spork 1998; Linke 1980). Critics point out that education *about* the environment (as traditional science or discipline-based learning) simply ignores important social aspects of human–environment relationships, while education *in* the environment simply changes the place in which traditional science learning occurs (Gayford 1996; Linke 1980). However, it is the notion of education *for* the environment that has caused the greatest consternation about the role of environmental education.

2.2.2 *Education* for *the Environment*

According to Stevenson (1987), education *for* the environment differs from education *about* and *in* the environment in terms of its goals, and the pedagogical approaches through which these goals are reached. He described education *for* the environment as working towards "socially critical and political action goals" (p. 69) through pedagogies that incorporate:

> the intellectual tasks of critical appraisal of environmental (and political) situations and the formulation of a moral code concerning such issues, as well as the development of a commitment to act on one's values by providing opportunities to participate actively in environmental improvement (p. 73).

This clearly positioned education *for* the environment as a critical, political endeavour, which aimed to promote and support the "transition to a socially just and ecologically sustainable society" (Fien 1993, p. 48). This means that for some critics, the term 'education *for* the environment' appears to contradict its intended goals. For example, N. Gough (1987) asserts that the term represents a "patronising and anthropocentric" perspective in that it objectifies human–environment relation-ships. He asked "who are we to say what is 'good' for the environment, and which environment is 'the environment' anyway?" (p. 50). He noted that the term supports "distinctions between subject and object, education and environment, learner and teacher" and therefore fails to be inclusive of alternative worldviews such as those representative of deep ecology in which humans see "themselves and nature as part of 'being'" (p. 50, original italics). "In order to shift our attention from the *objects* of environmental education" education must embrace an "ecological paradigm" that encourages students to "learn to live, and live to learn, *with* environments" (p. 50, original italics). However, N. Gough and A. Gough (2010) note that "learning *with* environments" requires a "radical socially critical pedagogy" that supports the "involvement of students in environmental action". Environmental education as "learning *with* environments" is "not yet common practice", in part due to the "timidity of many teachers and schools" to address the politically sensitivities of many environmental issues (p. 342, original italics).

This highlights the propensity for the term 'education *for* the environment' to be interpreted, or misinterpreted, in ways that reflect the preferred environmental and educational ideologies of the interpreter. Fig. 2.1 links the intention of education *about*, *in* and *for* the environment to specific environmental and educational ideolo-gies, the latter of which were derived from the work of Kemmis et al. (1983) and O'Riordan (1989), and which define major pedagogies and educational outcomes. The figure highlights modifications to Lucas' original terminology, suggested by authors attempting to locate components of environmental education within specific ideologies, including: education *from*, *through* and *with* the environment (Gough 1997), and conservative education *about* the environment, liberal education *about*, *through* and *for* the environment, and critical education *for* the environment (Fien 1993). According to Fien (1993), only an ecocentric, socially-critical approach to critical education *for* the environment fully addresses the intended goals of educa-tion *for* the environment described by Stevenson (1987) above. As such, a 'socially-critical education *for* the environment' demands an educational approach that supports "personal and social change" (Fien 1993, p. 49) as it aims to promote "eco-logically sustainable, people-environment relationship[s]" through "an overt agenda of political literacy, values education, and social change" (Thomas 2005, p. 108). This agenda has been the focus of much debate.

Socially-critical education *for* the environment has been labelled as overly deter-ministic by some critics, who believe it has the potential to indoctrinate students rather than facilitate the development of their own values and attitudes towards human–environment relationships (Jickling and Spork 1998; Burbules and Berk 1999). The notion that any educational practice can indoctrinate assumes that edu-cators are able to identify a specific "set of skills" and values or attitudes, that when taught, will lead to a specific behaviour (Scott and Gough 2003, p. 115). However, research regarding human constructed values, attitudes and beliefs, and their rela-tionship to human action, indicates that the premise that environmental education can teach specific or long-lasting environmental values or attitudes is unwarranted. Even altering an individual's value priorities is an extremely unlikely outcome,

		Educational ideology		
		Vocational neo-classical (prepare students for their future work)	**Liberal-progressive** (prepare students for their life in society)	**Socially-critical** (prepare students for their role in creating society)
Technocratic ↑	**Cornucopian** (environmental problems can be solved through science and technology)	**Conservative** education about the environment (environmental knowledge is obtained from positivist study of the natural sciences)		
Environmental ideology ↕	**Accommodation/ Managerialism** (environmental problems can be averted by good management of human–environment relationships)		**Liberal education** *about* **the environment** (environmental understanding is obtained through problem solving and enquiry-based study of the natural sciences)	
	Communalism/ Ecosocialism (cooperation will ensure that equality is part of all human-human and human–environment relationships)		**Liberal education** *in* (*through*) **the environment** (student-centred and experiential learning in environments outside the classroom)	**Critical/Socially-critical education** *for* (*with*) **the environment** (learning through decision-making, participation and action)
Ecocentric ↓	**Gaianism/ Utopian** (humanity is just one component of earth's natural systems, and is therefore subject to the same laws of nature)		**Liberal education** *for* **the environment** (identifying attitudes, values and beliefs through the case study of local environmental issues)	

Fig. 2.1 Educational and environmental ideologies in different approaches to environmental education (Adapted from Fien 1993, p. 40)

unless accompanied by significant and contextually specific experiences (Ajzen 1996; Fazio and Zanna 1981; Kraus 1995; Lewin and Grabbe 1945; Rokeach 1973).

Despite this, some educators prefer a liberal education *for* the environment to assist students to learn "how to think, not what to think" (Jickling 2003, p. 22). This has also been contested for naïvely assuming that it is possible to remove the influence of values and political agendas from educational endeavours. Huckle, for example, argued that this is not possible, and noted that values are "shaped by the material circumstance within which people live; circumstances sustained by powerful interests who can easily co-opt the ecological message and turn it to their advantage" (Huckle 1986, p. 6). Instead, the aim of socially-critical education *for* the environment is to assist students to recognise that people enact different values and value priorities in different contexts, and to provide opportunities through which students can "*derive* for themselves thoughts, actions and feelings" (Scott and Gough 2003, p. 115, original italics). Fien (1993) suggested that socially-critical education *for* the environment is best undertaken within a framework of "committed impartiality" which encourages teachers to "state rather than conceal their own views on controversial issues" and to "foster the pursuit of truth by insuring that competing perspectives receive a fair hearing through critical discourse" (Kelly 1986, p. 130). This approach positions learning not as "a process which acts on individuals' characteristics in order to change the world", but rather "one which challenges individuals' views of the world as a means of influencing their characteristics and hence ways of thinking and living" (Scott and Gough 2003, p. 119). This is not indoctrination.

It is important to note that Lucas' model places each of education *about, in* and *for* the environment as essential for holistic environmental education. This means that effective environmental education requires the deliberate inclusion and intent of education *for* the environment (Greenall 1980), not just within the science curriculum, but as an integral component of all learning activities (Linke 1980).

From this point on, the term 'education *for* the environment' refers to goals and practices consistent with the environmental and educational ideologies of a 'socially-critical education *for* the environment' discussed above, and as represented in Fig. 2.1.

2.2.3 Implementation of Lucas' Model

Lucas' (1972) notion of education *for* the environment was not without precedence, and had long been represented in schools outside Australia. For example, in Britain during the 1960s, school programs provided opportunities for students and communities to participate cooperatively in local environmental planning processes. By the 1980s the focus of this education had moved beyond local community concerns to embrace "the social use of nature and issues of environment and development at all scales" (Huckle 1991, p. 52). However, successive reviews of various environmental education programs and pedagogical practices in Australia (as discussed by Fien 1993), including a national evaluation (Linke 1980), a review by a study group of the Australian Curriculum Development Centre (CDC; Greenall 1980), case study evaluations undertaken as part of an CDC environmental education project (Robottom 1983; Stevenson 1986), and observations by Stapp and Stapp (1983) and Huckle (1987a, b), all reported the overwhelming absence of pedagogies supportive of the ideals and goals of education *for* the environment, even when these were appropriately expressed in curriculum guides. In other words, there were significant gaps between the rhetoric of education *for* the environment and the reality of teachers' practices.

2.3 Sustainable Development—A New Debate

As the public debate and concern about environmental issues continued to grow throughout the 1980s, understanding of human–environment relationships evolved to incorporate global perspectives and the complex interrelationships between the biophysical, social, economic and political aspects of any society (Fien 2001; Fien and Gough 2000). This encouraged the reconsideration of how to define and practice 'education *for* the environment', as reflected in recommendations presented in The World Conservation Strategy (IUCN 1980), the National Conservation Strategy for Australia (DHAE 1984), and the report of the World Commission on Environment and Development (WCED 1987). These reports considered the most critical goal for environmental education to be preparing societies to respond to twenty-first century challenges in ways that would maintain and preserve viable human–environment systems, and that in light of this, students must learn how to contribute to the development of sustainable societies (Fien 2001; Fien and Gough 2000; Gough 1997). The WCED suggested that "Education for Sustainable Development" (ESD) was an essential part of mitigating problems associated with increasingly complex human–environment relationships, noting that "'the world's teachers…have a crucial role to play' in helping to bring about 'the extensive social changes' needed for sustainable development to be achieved" (WCED 1987 quoted in Gough 1997, p. 32). This represented a significant change in environmental education discourse. ESD has become a strongly contested concept, both in terms of environmental ideology and its implications for the role of education in society (e.g. Fien 1993; Gough 1997; Scott and Gough 2003, 2004). It encompasses a broad range of concepts, "based on ideals and principles that underlie sustainability, such as intergenerational equity, gender equity, social tolerance, poverty alleviation, environmental preservation and restoration, natural resource conservation, and just and peaceable societies" (UNESCO 2005b, p. 28) which cannot be addressed by any single educational program.

The following discussion outlines the goals of ESD as represented in the documents that informed the curriculum and teachers' practices of the Australian Sustainable Schools Initiative (AuSSI). This program, an Australian Government initiative to implement ESD, focused on the "environmental preservation and restoration" and "natural resource conservation" (UNESCO 2005b, p. 28) components of ESD, hereafter referred to as environmental education.

Sustainable development has been described as a "shifting, indefinable and contingent concept" (Scott and Gough 2003, p. 125) founded on the future-oriented principle that the action of today's society "meets the needs of the present without compromising the ability of future generations to meet their own needs" (WCED 1987, p. 43). "But just what kind of sustainable development is education for sustainable development supposed to stand for?" (Kahn 2010, p. 16). Chapman (2004) noted that the term:

> sustainability, as it is employed in general usage, can mean anything you want. It has so many interpretations that it lacks any capacity to confront the reality of the unsustainable behaviour of our societies. The notions of sustainable growth, sustainable development and sustainable consumption (OCED 1999) link the concept of sustainability with language that has implicit meanings and assumptions that are technocratic and underlie the causes of environmental problems (p. 99).

Despite these inherent contradictory messages, ESD aims to embrace environmental education by "setting it in the broader context of socio-cultural factors and the socio-political issues of equity, poverty, democracy and quality of life" (UNESCO 2005a, p. 19), and is most significantly "about learning for change towards a more sustainable future" (Tilbury and Wortman 2004, p. 36). ESD places education not only as "a means of implementing" sustainable development (Scott and Gough 2003, p. 125), but also as an essential "part of a process of building an informed, concerned and active civil society" (Fien 2001, p. 17), through developing the "capacity of human beings to continuously adapt to their non-human environments by means of social organisation" (Hamm and Muttagi 1998, p. 2). These goals not only differ significantly from the common themes of Australian environmental education practices identified by Linke (1980), but also remain relatively abstract in terms of how they might be incorporated into educational practice (Scott and Gough 2003).

In 1992, the United Nations Conference on Environment and Development (UNCED) in Rio de Janeiro, Brazil, attempted to support "re-orientating education towards sustainable development" (Gough 1997, p. 33) through the establishment of twenty-seven sustainability principles incorporating key aspects of both environmental protection and human development. However, in the decade following the presentation of these principles, the establishment of ESD by schools, communities and governments was very slow (McKeown 2002), and there was a growing concern that globally, human–environment relationships were deteriorating at an ever-increasing rate (Gore 2006). In 2002, the United Nations World Summit on Sustainable Development (WSSD) in Johannesburg aimed to identify practical methods for implementing the sustainability principles established in Rio de Janeiro. In relation to education, the final Johannesburg Plan of Implementation (JPOI) stated that it was necessary to "Integrate sustainable development into education systems at all levels of education in order to promote education as a key agent for change" (WSSD 2002; Article 121).

In response to the WSSD recommendations, in 2002 the United Nations General Assembly proclaimed a Decade of Education *for* Sustainable Development (DESD) for the period 2005–2014 (WSSD 2002), outlining a vision for a future as "a world where everyone has the opportunity to benefit from education and learn the values, behaviour and lifestyles required for a sustainable future and for positive societal transformation" (DSE 2005, p. 4).

2.3.1   *Education* for *Sustainable Development (ESD)*

The notion of ESD as a vehicle for 'societal transformation' has created an opportunity to re-define the purpose and practice of education, but in so doing, presents an enormous challenge for educators. Although there is no agreed definition for what constitutes such transformative education, Morrell and O'Connor (2002) suggested that:

> transformative learning involves experiencing a deep, structural shift in the basic premises of thought, feelings, and actions. It is a shift of consciousness that dramatically and permanently alters our way of being in the world. Such a shift involves our understanding of ourselves and our self-locations; our relationships with other humans and with the natural world; our understanding of relations of power in interlocking structures of class, race and gender; our body-awareness, our visions of alternative approaches to living; and our sense of possibilities for social justice and personal joy (p.xvii).

This definition reveals the complexity and multiplicity of the inherent values, and the moral and ethical dimensions of the environmental and societal issues that position ESD as the precursor to action for social transformation towards sustainable development—expectations unlike any traditional subject, and beyond the capacity of the most pervasive or familiar teaching methods (Gayford 1996).

2.3.2   *Pedagogy for ESD*

As ESD "calls for additional and different processes than those traditionally thought of in education…to involve people, rather than convey just a body of knowledge" (Tilbury et al. 2002, p. 12), "issues of pedagogy are…vital in reorientating education towards sustainability" (Fien 2001, p. 23). However, "there is no absolute answer to the question of what is an appropriate pedagogical approach to learning in the context of sustainable development" (Scott and Gough 2004, p. 75). An effective pedagogy must not only encompass all of the scientific, technological, economic, aesthetic, political, ethical, cultural and spiritual aspects of human–environment interactions demanded by ESD, but also:

- inspire students' belief that they have the power and the responsibility to effect positive change on a global scale;
- encourage students to become primary agents of transformation towards sustainable development, increasing their capacity to transform their vision for society into reality;
- develop the values, behaviour and lifestyles required for a sustainable future;
- facilitate the learning of how to make decisions that consider the long-term future of the equity, economy and ecology of all communities; and
- build students' capacities for future-oriented thinking (AAEE 2005, p. 17).

Putting all of these into practice however, is problematic. Although it is evident that the acquisition of knowledge, often associated with traditional science education *about* the environment, does not fulfil the holistic aspirations of ESD, "the role of science and technology deserves highlighting as science provides people with the ways to understand the world and their role in it" (UNESCO 2005a, p. 18). In other words, science knowledge and environmental education should not be mutually exclusive (Gough 2007). Traditional science pedagogy however, conflicts with the behavioural outcomes of ESD, as transmissive, or vocational/neo-classical (Kemmis et al. 1983), teaching practices objectify the "biogeophysical" world, effectively separating humans from their environment and segregating facts from values (Scott and Gough 2004). As part of ESD, science pedagogy must incorporate more inclusive paradigms of teaching and learning to become oriented towards learning for action, or "science for action" (Gough 2007) in ways that "provide a scientific understanding of sustainability together with an understanding of the values, principles, and lifestyles that will lead to the transition to sustainable development" (UNESCO 2005a, p. 18). This reflects the understanding that holistic ESD must explore human activity as one part of the environment, and that this involves the role of human values and attitudes, or ideologies.

There is a long history of debate concerning the role of human values in environmental education (e.g. Lucas 1980). According to UNESCO, ESD is "fundamentally about values, with respect at the centre: respect for others, including those of the present and future generations, for difference and diversity, for the environment, for the resources of the planet we inhabit" (UNESCO 2005a, p. 6). Many human decisions and behaviours, including those related to the environment, are driven by values, value priorities, attitudes, and beliefs (Gayford 1996). This is the basis for recommendations for the incorporation of values education in ESD (e.g. the Belgrade Charter and Tbilisi Declaration), and is paralleled by studies indicating a pervasive belief amongst primary school teachers that environmental education must include the teaching of attitudes (Cutter and Smith 2001).

Values education however, is somewhat problematic. It requires educators to determine such things as what values are, how they are constructed, whose values should be taught, if values and attitudes can be actively learned, which learned values will cause a student to embrace a specific behaviour, and whether or not the teaching of values is simply indoctrination. Most importantly, educators must identify and assess the role of values embedded within the educational outcomes towards which they teach. This is particularly difficult when guiding statements, such as those that outline the role of ESD, contain apparently contradictory sets of values. For example, in Educating for the Future: A Transdisciplinary Vision for Concerted Action, UNESCO (1997b) states that "Sustainable consumption does not necessarily mean consuming less. It means changing unsustainable patterns of consumption by allowing consumers to enjoy a high quality of life by consuming differently" (quoted in Spring 2004, p. 121). For many, human consumer values are the root of today's environmental concerns, and yet this statement clearly retains the value of consumerism as a measure of life quality. Similarly, no single value has a universally agreed meaning or relative priority. For example, despite the development of

The Draft Strategy of Education for Sustainable Development in Sub-Saharan Africa in 2006, African educational institutions have been reluctant to embrace ESD. Manteaw (2012) attributes this, in part, to the belief that "meanings of sustainable development have been largely based on Western needs and values, which, to a large extent, have colonised local cultural interpretations and understandings. Additionally, the origins of the concept in global environment and development debates have given the concept an aura of 'globalness,' which, in many ways, is far removed from the day-to-day realities of local people" (p. 381).

Despite these issues, school communities in Australia do consider values education to be important, and identify the value of "individual responsibility" as essential (DSE 2005, p. 4), particularly as it relates to the maintenance and preservation of the environment (DEST 2005). There is, however, no definitive effective method for teaching 'individual responsibility'. The learning outcomes of any values education depends, in part, on the manner in which it is taught. Gayford (1996) notes that the behaviourist pedagogy employed in many environmental education classes may achieve little more than "green consumers", rather than developing the political literacy required to understand the role of values in the formation of complex and diverse societal environmental ideologies and resulting behaviours (McKeown 2002, p. 14). It is only through these understandings that environmental issues may be truly understood and "constructively resolved" (Clayton and Opotow 2003, p. 19). These outcomes require the use of a pedagogy that assists both teachers and students to begin to understand their own agency. Educators must understand the implicit political and social messages conveyed not only by the context of the content knowledge they teach, but equally also by the manner in which they teach it (Giroux 1997).

Effective ESD must therefore incorporate opportunities for developing understanding of human agency. This requires learning opportunities that facilitate students' understanding of the mechanisms of ideological conflict and resulting political forces, through critical examination of the past, present and potential future effects of human–environmental relationships. Teaching for social critique is therefore crucially concerned with facilitating understanding of how humans frame their ideas according to their values, attitudes and beliefs, how they construct their environmental ideologies and behavioural choices, and how these interact within a society (Scott and Gough 2003, 2004). The transformative learning outcomes of ESD are therefore necessarily associated with critical theory (Luke 2003).

2.3.3 Critical Theory

The notion of transformative learning, or transformative education, developed from the field of critical theory that originated during the 1920s at the Institute for Social Research in Frankfurt (Peters et al. 2003). The term 'critical theory' was coined by Horkheimer in 1937 to describe the philosophical and theoretical basis of work undertaken by the Frankfurt School, although the definition of the term changed and broadened over time (Peters et al. 2003). Although the early work centred on Marxist ideologies with the overriding goal to highlight the "critical function of Marxist theory as a form of opposition to bourgeois society" (Peters et al. 2003, p. 3), the focus of research broadened as new School members brought new perspectives. However, Horkheimer (1982) maintained a definition of critical theory that remains useful today: critical theory is related not to content, but to a philosophy directed mainly towards changing society in ways that "liberate human beings from the circumstances that enslave them" (p. 244). This definition incorporates the idea that "man [sic] can change reality, and the necessary conditions for such a change already exist" which implies that, unlike traditional positivist style outlooks on the world, humans are the "producers of their own historical way of life in its totality" (Peters et al. 2003, p. 3). Horkheimer valued the idea that humans are reflexive conscious beings, and that social reality is contextual (Horkheimer 1982; Horkheimer and Adorno 1972). It is this aspect or understanding of critical theory that informs the processes of transformative education identified as essential components of ESD.

2.3.4 Critical Theory as Transformative Education

Transformative education has been inconsistently related to various teaching practices and epistemological ideals, and various cultural and structural aspects of society (Schugurensky 2002). Although widespread use of the term emerged during the 1970s, there remains no single definition. The underlying principles of transformative education arose from a collection of ideas from many philosophers influenced by various social contexts, particularly the work of Paulo Freire, Antonio Gramsci and Karl Marx: "no education is politically neutral" as traditional education works to maintain the social status quo, particularly in relation to the overriding injustices or asymmetric power relations in society (Wink 2000, p. 77). This belief grew in response to an increasing awareness that social power asymmetries were defined and maintained not only by physical means, but also equally well by knowledge (Gramsci 1971), as "education is knowledge and knowledge is power" (Swain 2005, p. 1). Karl Marx for example, saw education as "an insidious vehicle for institutionalizing elite values and indoctrinating people into unconsciously maintaining" social power asymmetries (quoted in Wink 2000, p. 83). In light of this, emancipation (or transformation) was envisaged to begin with the development of critical awareness of the "social, economic and political dynamics of everyday situations and practices" (Schugurensky 2002, p. 61). This is the aim of critical pedagogy.

Critique, in terms of critical pedagogy, is about embracing critical perspectives. A common misconception is that critique is a negative process restricted to criticism; however, here it refers to a much deeper level of understanding that incorporates "seeing beyond" or finding new ways of understanding complexities, particularly in relation to self and the social world (Wink 2000, p. 29). The application of critical pedagogy however, does not guarantee that critique is holistic, or unaffected by the discourses through which it is practiced. Early practice of critical pedagogy reflected the prevailing "anthropocentric Marxist paradigm that assumes that humans are different from other species because of their ability to make choices" (Spring 2004, p. 132), and as such that nature is valued, understood and utilised only in terms of human needs (Bowers 1991). Similarly, much of the work of Habermas (1972, 1975) reflected values that placed nature in a "primal position prior to society" (Luke 2003, p. 239). Alternatively, critique conducted from a science-based positivist worldview may embrace Cartesian dualist views that objectify the environment, and which assume that issues relating to human–environment relationships may be assessed and/or categorised as either right or wrong (Bowers 1991). All of these are contrary to the reality of the social world where human action reflects a complex web of motivations and intentions, and contrary to desired ESD outcomes of holistically understanding the reality of dynamic and complex human–environment relationships. In the broadest sense, critical pedagogy acts as a pedagogy of transformation by teaching students to ask "for reasons why things are the way they are and why others (and oneself) act as they do" (Mogensen 1997, p. 430).

Since its inception, the notion of critical pedagogy has evolved in response to changes in society, and more recently, in relation to developing environmental perspectives. Before his death in 1996, Freire had begun to modify his ideas to incorporate environmental concerns, highlighting the need for a critical pedagogy he referred to as "ecopedagogy" (Spring 2004, p. 132), in order to critique the contribution of capitalist ideals to modern human–environment relationships. Freire's idea inspired many pedagogical developments. Gadotti (1994), for example, built upon this idea to define "planetary consciousness" as a more holistic alternative pedagogical focus (quoted in Spring 2004, p. 133), and Kahn (2010) presented ecopedagogy as the basis for a holistic framework for ESD:

> Ecopedagogy seeks to interpolate quintessentially Freirian aims of the humanization of experience and the achievement of a just and free world with a future oriented ecological politics that militantly opposes the globalization of neoliberalism and imperialism, on the one hand, and attempts to foment collective ecoliteracy and realize culturally relevant forms of knowledge grounded in normative concepts such as sustainability, planetarity, and biophilia, on the other (p. 18).

Irrespective of the intended focus or ultimate aim of any form of critical pedagogy, the practical application of pedagogy determines its effectiveness. The understanding that the most effective critical pedagogy encompasses understandings unique to a place and time became known as socially-critical pedagogy (Giroux 1988).

2.3.5   *Socially-Critical Pedagogy* **for** *Learning*

The notion of socially-critical pedagogy was founded on the understanding that learning is only truly effective when developed within contexts related to a student's life experiences (Giroux 1988)—that is, within their "community" (Mogensen

1997, p. 434). Socially-critical pedagogy deliberately and specifically deconstructs political, social and economic motivations for human action, thereby providing commentary on human values, value priorities, attitudes and beliefs (Fien 1993). As this pedagogy engages students in considering the complexity and dynamics of such human ideas, it supports the outcomes of ESD as "it is action on the basis of comprehensive reflection which decisively changes the conditions of human life" (Mogensen 1997, p. 431).

The effectiveness of a socially-critical pedagogy is also dependent upon the manner in which students partake in such significant and contextually specific experiences. This is highlighted by Freire's (1972) early work in which he identified two main educational forms with opposing relationships between power and school education—"banking" and "liberation" education—where students are positioned as either a "passive subject" or "active actor" respectively (Swain 2005, p. 1). The role of the learner as an active actor is central to a socially-critical pedagogy. Although critical pedagogy in general was seen to provide opportunities for developing awareness and engaging in effective critical reflection, Freire believed that this would be truly transformative only if accompanied by social action, or authentic participation (Schugurensky 2002, p. 63). In many ways this reflects Lucas' (1979) idea that learning about sustainable human–environment relationships from others does not necessarily lead to similar action. Transformative learning, or learning that empowers individuals to participate in the development of sustainable human–environment relationships, comes only from direct participation in these behaviours. In other words, socially-critical education *for* the environment encourages learning through:

> just, participatory and collaborative decision making, and involves critical analysis of the development of the nature, forms and formative processes of society generally and of the power relationships within a particular society, thus revealing how the world works and how it might be changed (Gough 1997, p. 107).

Similarly, Gruenewald (2003) proposed a "critical pedagogy of place" as an approach which draws upon the ideals of both critical pedagogy and place-based education to contextualise education in ways that enable students to "interrogate the intersection between cultures and ecosystems" (p. 10) so that it has a "direct bearing on the well-being of the social and ecological places people actually inhabit" (p. 3). In addition, if ESD through a socially-critical pedagogy is to be most effective, Schugurensky (2002) points out that student participation must be legitimately incorporated throughout the organisational structures of their schools, as:

> when people have the opportunity to actively participate in deliberation and decision making in the institutions that have most impact on their everyday lives, they engage in substantive learning and can experience both incremental and sudden transformations. The transformative effects are usually more significant when this institutional participation provides empowering experiences (p. 67).

Freire (1994) believed that in the absence of authentic participation, a socially-critical pedagogy not only failed to lead to behavioural change, but also actively discouraged such change.

Critical reflection, without an accompanying effort of a social organisation and without concurrent enabling structures to channel participation in democratic institutions, can nurture the development of individuals who become more enlightened than before, but who (because of their realisation of the immense power of oppressive structures) may become more passive and skeptical than before (Schugurensky 2002, p. 62). This effect may be caused by a tendency of social critique, in the absence of authentic participation, to emphasise negative relationships which contribute to student despair and feelings of being unable to influence their world. It is therefore essential that students are engaged in positive or "empathetic and optimistic" reflection orientated towards solutions to which they can personally contribute (Breiting et al. 2005). This is supported by John Dewey's ideas that democracy as an ideology cannot simply be studied, but must be lived to be understood (Wink 2000), and that this lived experience must be accompanied by a "language of possibility"—a belief that as an individual there are opportunities for positive change (Fien 1993, p. 10). In other words, effective learning through a socially-critical pedagogy depends on the manner in which teachers implement it.

2.3.6 Socially-Critical Pedagogy and Teachers

In order to best achieve the outcomes of social transformation through a socially-critical pedagogy, Gramsci (1971) noted that educators must first "recognise and acknowledge the existing oppressive structures inherent in schools" in order to actively empower learners to change "beliefs into behaviours for self and social transformation" (quoted in Wink 2000, pp. 82, 85). In other words, transformative education, or the ideals of transformative learning, requires educational processes to change from indoctrinating learners into accepting existing social structures, to empowering learners to actively shape, or indeed re-shape, their society. Both educators and learners are integral to the transformative process undertaken through a critical pedagogy as:

> a way of thinking about negotiating and transforming the relationship among classroom teaching, the production of knowledge, the institutional structures of the school, and the social and material relations of the wider community, society, and nation state (McLaren 1998, p. 48).

This, however, is an enormous undertaking. "It is a very strong indictment to say that our conventional educational institutions are defunct and bereft of understanding of our present planetary crisis" and "transformative education fundamentally questions the wisdom of all current educational ventures" (O'Sullivan et al. 2002, p. 10). In other words, the practice of a socially-critical pedagogy, as transformative ESD, is a radical process. It requires educators to question their current educational practices and the broader practices of the society to which they contribute in order to build the capacity of their students to reflect critically on the predominant human–environmental relationships that support, and are supported by their society. In order to embrace ESD, educators must actively challenge the predominant political values from which today's "relentless and expansive exploitation of nature" and the underlying notion that equality is a measure of equal access to consumer goods has evolved (Luke 2003, p. 239). They must find ways to re-direct the current economic and consumerist educational outcomes to goals that are more aligned with sustainable development. All of these actions require educators to challenge existing human-centred ideals with educational theories and practices that view human life as an integral component of Earth's natural systems (Spring 2004).

Socially-critical education implies dissatisfaction with current dominant social paradigms, many of which may be directly threatened by critical appraisal of their environmental ideologies. However, in a democratic society, the notion of educating for a specific type of social transformation, even with agreement regarding the types of transformation desired, understandably attracts concern.

2.4   Development of Socially-Critical ESD in Australia

The development of environmental education in Australian schools, in terms of both policy development and classroom practice, has been well documented by Fien (1993) and Gough (1997). By the late 1990s, Australian educational agencies began to re-consider their roles and responsibilities in defining and implementing environmental education in light of the developing notion of ESD. In 1999, the Ministerial Council on Education, Employment, Training and Youth Affairs (MCEETYA) acknowledged the importance of environmental education as Goal 1.7 of The Adelaide Declaration on National Goals for Schooling in the Twenty First Century:

> Schooling should develop fully the talents and capacities of all students. In particular when students leave school they should have an understanding of, and concern for, stewardship of the natural environment, and the knowledge and skills to contribute to ecologically sustainable development (MCEETYA 1999, p. 1).

In 1999, the Department of the Environment and Heritage (DEH) established an educational reference group to explore ways in which Australian schools should respond to the United Nations Agenda 21 framework for environmental education. Their discussion paper, Today Shapes Tomorrow: Environmental Education for a Sustainable Future, defined environmental education as:

- raising awareness;
- acquiring new perspectives, values, knowledge and skills; and
- formal and informal processes leading to changed behaviour in support of a sustainable environment (DEH 1999, p. 4).

The paper noted that, despite the government rhetoric advocating sustainable development, "actions have failed to adequately reflect these commitments to environmental education" (DEH 1999, p. 22), as environmental education was isolated within schools and focused towards knowledge acquisition and attitudinal change.

They concluded that effective education *for* sustainability required "comprehensive, lifelong environmental learning integrated within education systems, industry, social organizations/neighbourhood groups and government" because the "transition from awareness to knowledge and action must be owned by all" (DEH 1999, p. 22).

This paper informed the Australian Government's Environmental Education for a Sustainable Future: National Action Plan, which was launched in 2000 as the "starting point for an enhanced national effort in support of Australia's ecologically sustainable development" (DEH 2000, p. 3). This plan acknowledged that environmental education must: involve everyone; be lifelong; be holistic and about connections; be practical; and be in harmony with, and of equal priority to, other social and economic goals (DEH 2000). Although the action plan was not intended to be a definitive model for environmental education, several important aspects of the earlier discussion paper were poorly represented, typified by the statement that a key element of environmental education "is a move from an emphasis on awareness raising to an emphasis on providing people with the knowledge, values and skills to actually make a difference to the protection and conservation of the Australian environment" (DEH 2000, p. 3). This outdated notion of environmental education embraced a parochial view of local conservation rather than a global perspective, and associated education with the delivery of appropriate ideas, or values, as instigating effective behavioural change. The role of knowledge acquisition was somewhat qualified by the statement: "Specialist discipline-based knowledge, while contributing critically, is no longer adequate by itself—an holistic appreciation of the context of environmental problems is essential" (DEH 2000, p. 4). In other words, the base line for evaluating good environmental education continued to be associated primarily with the acquisition of knowledge and understanding, rather than by outcomes evidenced by individuals' actions.

A critical element of the Action Plan was the establishment of several non-statutory bodies to initiate, monitor and evaluate environmental educational initiatives, provide expert advice to government, and collaborate to develop a national approach for environmental education presented as the National Environmental Education Statement for Australian Schools—Educating for a Sustainable Future. This statement, endorsed by the Ministerial Council on Education, Employment, Training and Youth Affairs (MCEETYA), represented the first national approach to environmental education to be endorsed by all Australian federal, state and territory governments, and reflected the growing understanding at the time that effective environmental education was indeed a priority (DEH 2005).

Although this statement generally supported the visions and sentiments of environmental education outlined in preceding Australian Government documents, it succeeded in more comprehensively highlighting the global and holistic characteristics of environmental education by relating it to the "interdependence of social, cultural, economic and ecological dimensions at local, national and global levels" (DEH 2005, p. 8). Most importantly, the statement directly acknowledged "action and participation" as essential outcomes of environmental education, and indicated

(although did not state specifically) that changes towards a socially-critical pedagogy were desired. The educational "vision" for students was that they become
"active, self-directed and collaborative learners and ethical and responsible citizens
taking action for a sustainable future" (DEH 2005, p. 8) by developing:

- a willingness to examine and change personal lifestyles to secure a sustainable
  future;
- the ability to identify, investigate, evaluate and undertake appropriate action to
  maintain, protect and enhance local and global environments;
- a willingness to challenge preconceived ideas, accept change and acknowledge
  uncertainty; and
- the ability to work cooperatively and in partnership with others (DEH 2005,
  p. 10).

The vision for teachers similarly hinted at a need for change, as they were to
become "enthusiastic about teaching and about developing effective relationships
with their students, committed to the goals of education *for* sustainability, life-long
learners, adaptable, and open to new ideas and teaching strategies" (DEH 2005,
p. 8). However, the document contained mixed messages about how such 'visions'
for environmental education should be incorporated in classroom practices. The
most direct reference to a socially-critical pedagogy for environmental education
was reflected by the understanding that:

> An environmental education for sustainability curriculum involves understanding the pres-
> ent—how it has been shaped, the value in which it is held, and seeking to mitigate adverse
> effects on it. This involves an investigation of how we have come to this situation and
> accepting responsibility to work towards a sustainable future (DEH 2005, p. 13).

The suggested teaching strategy for this is outlined as an inquiry learning model
incorporating experiential learning and science in the community. In a move away
from a traditional vocational/neo-classical pedagogy, learning through social action
is encouraged through a requirement that "students be active in decision making
during the inquiry and at its conclusion" (DEH 2005, p. 21).

In 2007, the Australian Government presented a national strategy for fostering
sustainable development through environmental education: Caring for Our Future—
The Australian Government's Strategy for the United Nations Decade of Education
for Sustainable Development, 2005–2014 (DEH 2007). This strategy stated that
"the Australian community will have the understanding, knowledge, skills and
capacity to contribute to sustainable development and will embrace the intrinsic
value of sustainability as a national aspiration" (DEH 2007, p. 4) but provided little
evidence of encouraging actual action, or guidelines for how this should be achieved.
In terms of "communicating the concepts" (DEH 2007, p. 5) of sustainable development, the strategy highlighted the need to foster collaborative partnerships between
government, business and community, and supported the Australian Sustainable
Schools Initiative (AuSSI) as one program through which this could be achieved.

2.4.1 The Australian Sustainable Schools Initiative, Victoria

In 2001 the Sustainable Schools Working Group was established to oversee the development and implementation of what was to become the Australian Sustainable Schools Initiative (AuSSI), an Australian Government initiative which aimed to assist schools and communities to move towards environmental sustainability by facilitating authentic co-learning opportunities as part of a whole-school approach to environmental education—in essence, to develop socially-critical ESD. In 2003, the AuSSI initiative began as an 18 month pilot study during which 300 schools across Victoria and New South Wales began to implement the Sustainable Schools Program (SSP).

In Victoria, 113 schools participated in the pilot study. SSP was funded jointly by the Commonwealth Department of Environment and Heritage (DEH) and the Victorian Department of Education and Training (DET), and delivered by the Gould League and the Centre for Education and Research in Environmental Strategies (CERES). Facilitators from the Gould League and CERES assisted schools with implementation issues, provided teacher professional development and liaised closely with in-school SSP coordinators. This high level of support was crucial because, at this time, environmental education was not mandatory in Victoria, and in many schools, neither teachers, nor students, were familiar with basic environmental concepts (Larri 2006).

2.4.2 Aims of the Sustainable Schools Program

The Sustainable Schools Program was developed to translate into effective educational practice the critical elements of government documents and statements which advocated environmental education as the essential precursor to sustainable development. The program was predicated on several key understandings that had been poorly expressed in education policies. The most important of these was the understanding that building awareness of environmental issues does not necessarily predict the willingness or ability of people to undertake pro-environmental behaviour (Hungerford and Volk 1990), because "there is often little or no relationship between attitudes and or knowledge and behaviour" (McKenzie-Mohr and Smith 1999, p. 10). In other words, there was a growing understanding that effective education *for* the environment or *for* sustainable development depended not so much on what was taught, but on how it was taught. SSP positioned schools as communities which modelled environmental sustainability—places in which environmental learning embraced collaborative ventures which contributed directly to the sustainable operation of the school and community. Table 2.1 shows that the program consisted of twelve steps that aimed to facilitate a school's journey from awareness to action in a manner that brought with them not only the teachers and students, but also their local community.

Table 2.1 The twelve key elements of the framework for facilitation of the Sustainable Schools Program

Key Element	Why this element is important
Introduction to sustainability	Provides a vision, unity, an understanding of the issues and a broad plan for the future. Without this introduction, there will be no common purpose or vision.
Collect baseline data	Provides key information against which future change can be measured. Provides a reference point to track progress.
Make a whole school commitment	A commitment from all sectors of the school to become more sustainable is crucial for a whole school change. Ensures change will develop beyond isolated pockets in the school, breaks down resistance.
Form a committee	A committee, with representatives drawn from teachers, parents, students and specialist advisors, will give ownership to all sectors in the school and a structure to ensure that the workload is spread over the group. A committee shares the load among dedicated teachers and provides ownership by the rest of the school.
Conduct an assessment / audit	Assessment and audits can give reliable information on how resources are used in a school and how waste and litter is being generated. A plan provides certainty.
Set goals and targets	By setting goals and targets, a school will focus on achieving measureable outcomes with clear direction.
Develop a policy	A policy embeds a programme in a school, gives the programme long-term approval.
Develop action plans	Action plans provide a structure and a sense of organisation to achieve outcomes.
Develop curriculum plans	Curriculum plans identify where sustainability is being covered in the school's curriculum and set an operationally coordinated approach.
Implement actions and curriculum plans	Implementation is the essential and exciting step for staff and students.
Monitor and evaluate the programme	Monitoring and evaluation assists a school to constantly re-evaluate its effectiveness and provide constant improvement in their programme.
Build community links	Community links enrich a school's programme bringing valuable resources, expertise and support to and from their wider community.

Larri (2006, p. 20)

Note: This table is an excerpt of documentation provided by the Gould League to the Victorian Department of Education and Training to describe their approach to the Sustainable Schools Program

Schools undertaking SSP began by implementing a core module of activities designed primarily to raise awareness within the school and school community, and to collect data regarding the resource usage of the school. This data informed the development of a plan to implement sustainable school management and operational policies, centred around four resource-based modules: water, waste, energy and biodiversity. The aim of the initial stages of SSP was to "foster school ownership and empowerment of their sustainability program with a focus on student involvement and learning" (Larri 2006, p. 3). Table 2.2 shows the conceptual model

Table 2.2 Conceptual model for the Sustainable Schools Program

Level 8	Schools are working as models of sustainability in their communities		Ultimate impacts	Intended longer term impacts of the AuSSI Pilot – criteria for success not yet clearly delineated
Level 7	7a. Active and empowered students continuously work towards sustainability	PERSONAL RESPONSIBILITY		
	7b. Whole school change management is underpinned by decisions that work towards sustainability	SCHOOL SUSTAINABILITY		
	7c. Changes in the wider community are based on decisions that work towards sustainability	COMMUNITY SUSTAINABILITY		
Level 6	Schools monitor and evaluate their plans, review and modify them as required, record benefits (environmental, economic, educational and social), and celebrate and build on their achievements		Resource themes / modules	Successive completion of each resource based module or theme through: a process of policy development, action planning, and curriculum planning; and then implementation of action plans, monitoring, review, and celebration of achievements
Level 5	For each resource theme, schools implement the action plans and curriculum plan, and build links to the wider community			
Level 4	For each resource theme, schools develop a policy, action plans, and curriculum plans			
Level 3	Schools make a whole school commitment to become a Sustainable School, with long-term goals and targets for operations, curriculum, and whole school engagement across four resource themes		Core module	Completion of the core module "Becoming a Sustainable School" is equivalent to achieving Levels 1, 2 and 3. This is documented by the 4-Year Plan which includes: the baseline data set and results of curriculum audits; an agreed school vision; and sub-strategies to achieve action in each of the resource areas (i.e. waste, water, energy, and biodiversity)
Level 2	School communities develop a deeper understanding of what it means to live sustainably and a shared vision of their school as a Sustainable School			
Level 1	Schools are aware of their current situation and identify the drivers and barriers to becoming a Sustainable School			

Larri (2006, p. 23)

Note: This outcomes hierarchy was developed as part of a larger comparative evaluation of both NSW and Victorian Sustainable Schools pilots. The evaluation of the NSW pilot was in collaboration with Sue Funnell, and this outcomes hierarchy was developed from the NSW outcomes hierarchy (See Larri 2006, pp. 19–23) upon which SSP modules were based, and through which it was hoped that schools would progress to become "working models of sustainability in their communities" (Larri 2006, p. 23).

Larri (2006) reported that Victorian schools participating in the SSP pilot study viewed the program as "an wholistic approach to our environmental management and sustainability programme and its integration into teaching and learning" (p. 42). They believed that the program would be easy to implement because it "provided a mechanism for managing change by providing structure, direction and momentum" (p. 40). They also valued the associated accreditation scheme which formally acknowledged and rewarded schools for the completion of each module, and was seen as a way in which to increase community awareness of the environment and schools' engagement with sustainability issues (Larri 2006).

In an evaluation of one aspect of the pilot SSP implementation (the Stormwater Action Project) in six Victorian schools, the success of the program was attributed to the "shared vision of teachers, students and parents that the environment has a high profile in the school" (Gough 2004, p. 29). Schools reported a wide variety of "educational benefits for students, social benefits for the whole school community, and professional benefits for teachers" (Larri 2006, p. 36). The core units of the program assisted teachers with "understanding the issues around sustainability" (Gough 2004, p. 29), and the teachers valued the opportunities to engage and learn with others (Larri 2006). Teachers noted that the whole-school approach effectively encouraged their students to become involved in environmental decision making processes while adequately accommodating all students' learning needs and interests. This increased the students' understanding and engagement in sustainability issues and motivated them to assume greater personal responsibility for their actions, as evidenced by reports that many students had initiated changes in their homes. In other words, the implementation of SSP achieved behavioural change towards sustainable practices within the schools and the wider community. The schools also reported that changes made in response to the initial resource auditing module provided significant resource and monitory savings, the latter of which were often reinvested into environmental education resources and activities. The majority of the schools indicated that changes implemented through SSP, particularly those related to the routine usage of resources such as water and energy, appropriate management of waste, and the maintenance of new equipment such as rainwater tanks, would prevail for at least a year (Gough 2004; Larri 2006).

Although these reports indicated that the implementation of SSP was successful in achieving some critical environmental educational aims, other reports can be interpreted to indicate that some of these changes were temporary. Many schools felt that SSP facilitators did not always understand or appreciate the operational issues or the difficulties faced by schools trying to implement change. Despite this, most of the schools were concerned that SSP facilitators were not a permanent resource (Larri 2006). This implies that, although the core modules aimed to assist the schools to develop ownership of the change process, not all of the schools had achieved a state of confidence or self-sufficiency in their journey towards becoming more sustainable.

2.5   The Environmental Educational Rhetoric–Reality Gap

Despite consistent calls for ESD for many years now, uptake of effective ESD in Australian educational policy and classroom practice has been slow (Fien 2001; Tilbury et al. 2004). This illustrates the common observation that teaching practices have an inertia that is difficult to shift (Fullan 2007; Hargreaves 1997; Scott and Gough 2003). As noted by Donnison (2004), "teachers and educational institutions are resistant to change" (p. 26), in part, because "the way that teachers are trained, the way that schools are organised, [and] the way that the educational hierarchy operates…results in a system that is more likely to retain the *status quo* than to change" (Fullan 2003, p. 3, original italics). The "lack of coherence between learning objectives and the practice of teaching" (Sørensen 1997, p. 179), is referred to as an educational rhetoric–reality gap (Stevenson 1987, 2007a).

Environmental education rhetoric–reality gaps have been an observed phenomenon in Australian schools since the first calls for environmental education to depart from traditional science, knowledge-based instruction during the 1970s. An extensive investigation by Stapp and Stapp (1983) of the status of education *for* the environment in Australia during 1982 revealed significant rhetoric–reality gaps. They reported that at this time, teachers' practices: were not "interdisciplinary"; did not provide opportunities for "problem solving"; avoided controversial issues which required confronting "values"; and failed to place learning in outdoor or real world contexts. In general, teachers viewed the environment as "nature", excluding important human–environment relationships of the more "urban" regions which represented most students' "own local environment". This investigation concluded that teachers tended to act as "conveyors of information, not facilitators" with a "strong emphasis in the higher grades on academic achievement" (Stapp and Stapp 1983, p. 5). In 1984, a similar study concluded that Australian educators taught in a manner in which the environment was "portrayed as somewhere where people do not live. The focus is on the natural and the nice and not connected at all with the everyday real experiences of living in towns or cities" (Bishop and Russell 1985, p. 14). Such observations are not restricted to environmental education in Australia, nor just to the earliest attempts to introduce education *for* the environment. Despite over 40 years of calls for practices in schools to depart from a knowledge-based vocational/neo-classical pedagogy in order to accommodate the goals of education *for* the environment through socially-critical pedagogies, traditional vocational/neo-classical pedagogies remain predominant (McKeown 2002). Eilam and Trop (2011) noted that "Although the contents of learning have changed, the prevailing pedagogy is still the same as it was throughout the 100 years in which the environmental crisis was developing" (p. 43).

The development of such educational rhetoric–reality gaps is not unexpected, due to the demands of the socially-critical and transformative educational goals of education *for* the environment, and more recently, ESD (Bishop and Russell 1985; Fien 2001; Robertson and Krugly-Smolska 1997; Stapp and Stapp 1983; Stevenson 2007b). Embracing socially-critical pedagogies requires educators and institutions to alter the well-established ways of thinking that have not only underpinned the educational routines that traditionally act to reproduce current human–environment relationships, but which also ideologically and practically contradict ESD outcomes (Kemmis 1991).

In other words, environmental education programs, and the social and cultural discourses embraced by socially-critical pedagogies, are inherently political such that "if properly implemented, they could be most threatening" (Greenall 1987, p. 13) for teachers, particularly during instances of conflict between their own views and those presented by the school, students and their families (Linke 1984). In light of the challenge of such a significant change, Scott and Oulton (1999) noted that teachers and schools have been poorly guided by "a bewildering mixture of often contradictory instruction", particularly in terms of maintaining a traditional academic assessment process while implementing learning that addresses the socially-critical, transformative goals for sustainable development (p. 90). There is generally a "lack of clear guidelines regarding EE/ESD pedagogy that contributes to this ambiguity and lag between practice and rhetoric" (Eilam and Trop 2011, p. 56). Many teachers do not believe that they either have the expertise to undertake such teaching, or that it is their responsibility to do so (Fien 1993).

More than anything else, the long history of observed rhetoric–reality gaps in the implementation of ESD suggests that the theory of environmental education is "not sufficiently grounded in teachers' experiences and in what they feel schools can do, or what the school day is really like" (Robertson and Krugly-Smolska 1997, p. 232). This has led to such rhetoric–reality gaps being attributed to myriad causes, including deficient teacher training, insufficient teacher knowledge, and a lack of time and school resources (e.g. Barrett 2007; Chapman 2004; Fien 1993; Grace and Sharp 2000; Spork 1992; Thomas 2005; Vongalis-Macrow 2007). A socially-critical approach to ESD is often viewed as impractical in that it not only fails to provide teachers with an "implementation" framework, but also "denies their own practical knowledge" (Walker 1997, p. 5). Stevenson (2007a) however, predicts that despite the "substantial" rhetoric–reality gap in environmental education, with increased dialogue and "research for addressing the gap", the "possibilities for enacting critical and substantive environmental education practices in schools" can be identified (p. 137), particularly if the rhetoric–reality gap is reconceptualised so that "practices in schools are not simply assessed in relation to policy discourse but policy discourse itself is re-examined in relation to teachers' practical theories and the contexts shaping their practices" (p. 265). Thus, there remains "a need to provide updated information on many aspects of environmental education in the school curriculum to inform policies for curriculum development and teacher education" (Lee and Williams 2001, p. 218).

2.6 Moving Forward

The documents from which this brief history of the development of ESD as effective education *for* the environment was compiled focused almost entirely on desired educational outcomes. Embedded within the outcome statements of these documents were the assumptions that not only could ESD learning outcomes be pre-determined, but that students would also embrace ESD and actively respond to what they learned. In light of these assumptions, programs such as SSP endorsed a socially-critical pedagogy as the most appropriate classroom approach to the goals of this socially transformative education. However, as these documents failed to indicate how the practice of ESD relates to the ontology of the educational environments in which it is implemented, it is difficult to assess the relationship between the stated ESD outcomes and student learning, or the appropriateness of a socially-critical pedagogy. As a result, the implementation of ESD programs often result in the development of educational rhetoric–reality gaps. In order to find ways in which to more effectively implement ESD, it is essential to understand the educational environments and pedagogical practices through which ESD outcomes are to be achieved. Chapter 3 introduces Anthony Giddens' theory of structuration as an ontological framework that outlines and explains the complexity and dynamics of the social interactions that constitute an educational institution, and that can effectively inform investigations into the development of rhetoric–reality gaps in the practices of teachers implementing ESD programs such as SSP.

References

AAEE. (2005). *United Nations decade of education for sustainable development 2005–2014. A proposal for action*. Melbourne: Australian Association for Environmental Education.

Ajzen, I. (1996). The directive influence of attitudes on behavior. In P. M. Gollwitzer & J. A. Bargh (Eds.), *The psychology of action: Linking cognition and motivation to behavior* (pp. 385–403). New York: Guilford Press.

Barrett, M. J. (2007). Homework and fieldwork: Investigations into the rhetoric-reality gap in environmental education research and pedagogy. *Environmental Education Research, 13*(2), 209–223.

Bishop, J., & Russell, G. (1985). Study tours in Australia and the USA. *Review of Environmental Education Developments, 13*(2), 14–15.

Bowers, C. A. (1991). The anthropocentric foundations of educational liberalism: Some critical concerns. *The Trumpeter, 8*(3), 102–107.

Breiting, S., Meyer, M., & Mogensen, F. (2005). *Quality criteria to develop ESD-schools: Guidelines to enhance the quality of education for sustainable development*. Vienna: Austrian Federal Ministry of Education, Science and Culture.

Burbules, N. C., & Berk, R. (1999). Critical thinking and critical pedagogies: Relations, differences, and limits. In S. P. Thomas & L. Fendler (Eds.), *Changing terrains of knowledge and politics* (pp. 45–65). New York: Routledge.

Chapman, D. (2004). Sustainability and our cultural myths. *Canadian Journal of Environmental Education, 9*, 92–108.

Clayton, S., & Opotow, S. (2003). *Identity and the natural environment: The psychological significance of nature*. Cambridge: MIT Press.

Cutter, A., & Smith, R. (2001). Gauging primary school teachers' environmental literacy: An issue of priority. *Asia Pacific Education Review, 2*(2), 45–60.

DEH. (1999). *Today shapes tomorrow: Environmental education for a sustainable future – A discussion paper*. Canberra: Environment Australia, Department of Environment and Heritage.

http://www.environment.gov.au/archive/education/publications/discpaper/index.html. Accessed 30 Sept 2014.

DEH. (2000). *Environmental education for a sustainable future*. Canberra: Department of the Environment and Heritage. http://www.environment.gov.au/system/files/resources/c4a11a84-8d2b-4abe-983e-e337e9f870f4/files/nap.pdf. Accessed 30 Sept 2014.

DEH. (2005). *Educating for a sustainable future: A national environment education statement for Australian schools*. Canberra: Department of the Environment and Heritage. http://www.environment.gov.au/system/files/resources/1b93d012-6dfb-4ceb-a37f-209a27dca0e0/files/sustainable-future.pdf. Accessed 30 Sept 2014.

DEH. (2007). *Caring for our future: The Australian government strategy for the United Nations Decade of Education for Sustainable Development, 2005–2014*. Canberra: Department of the Environment and Heritage. http://www.environment.gov.au/resource/caring-our-future-australian-government-strategy-united-nations-decade-2005-2014. Accessed 30 Sept 2014.

DEST. (2005). *National framework for values education in Australian schools*. Canberra: Department of Education, Science and Training. http://www.curriculum.edu.au/verve/_resources/Framework_PDF_version_for_the_web.pdf. Accessed 30 Sept 2014.

DHAE. (1984). *A national conservation strategy for Australia*. Canberra: Department of Home Affairs and Environment.

Donnison, S. (2004). The 'digital generation', technology, and educational change: An uncommon vision. In B. Bartlett, F. Bryer, & D. Roebuck (Eds.), *Educating: Weaving research into practice* (Vol. 2, pp. 22–31). Brisbane: School of Cognition, Language and Special Education, Griffith University.

DSE. (2005). *National framework for values education in Australian schools*. Canberra: Department of Science and Education. http://www.curriculum.edu.au/verve/_resources/framework_pdf_version_for_the_web.pdf. Accessed 1 Oct 2014.

Eilam, E., & Trop, T. (2011). ESD pedagogy: A guide for the perplexed. *The Journal of Environmental Education, 42*(1), 43–64.

Fazio, R. H., & Zanna, M. P. (1981). Direct experience and attitude-behavior consistency. In L. Berkowitz (Ed.), *Advances in experimental social psychology* (Vol. 14, pp. 161–202). New York: Academic.

Fien, J. (1993). *Education for the environment: Critical curriculum theorising and environmental education*. Geelong: Deakin University Press.

Fien, J. (2001). *Education for sustainability: Reorientating Australian schools for a sustainable future* (Tela series). Fitzroy: Australian Conservation Foundation. http://www.acfonline.org.au/sites/default/files/resources/tela08_education_%20for_sustainability.pdf. Accessed 30 Sept 2014.

Fien, J., & Gough, A. (2000). Environmental education. In R. Gilbert (Ed.), *Studying society and environment. A handbook for teachers* (pp. 200–216). South Melbourne: Macmillan.

Freire, P. (1972). *Pedagogy of the oppressed*. Harmondsworth: Penguin.

Freire, P. (1994). *Pedagogy of hope*. London: Continuum.

Fullan, M. (2003). *Change forces: Probing the depths of educational reform* (3rd ed.). London: Falmer.

Fullan, M. (2007). *The new meaning of educational change* (4th ed.). New York: Teachers College Press.

Gadotti, M. (1994). *Reading Paulo Freire: His life and work*. Albany: State University of New York Press.

Gayford, C. (1996). The nature and purpose of environmental education. In G. Harris & C. Blackwell (Eds.), *Environmental issues in education* (pp. 1–20). Hants: Ashgate Publishing.

Giroux, H. A. (1988). *Teachers as intellectuals*. New York: Bergin and Garvey.

Giroux, H. A. (1997). *Pedagogy and the politics of hope: Theory, culture and schooling*. Colorado: Westview Press.

Gore, A. (2006). *An inconvenient truth. The planetary emergency of global warming and what we can do about it*. London: Bloomsbury.

Gough, A. (1997). *Education and the environment: Policy, trends and the problems of marginalisation.* Melbourne: Australian Council for Educational Research.

Gough, A. (2004). *Evaluation of the sustainable schools stormwater action project 2003/2004 for the Gould League.* Burwood: Deakin University.

Gough, A. (2007, November 25–29). *Beyond convergence: Reconstructing science/environmental education for mutual benefit.* In Annual conference of the Australian Association for Research in Education, Freemantle, Australia.

Gough, N. (1987). Learning with environments: Towards an ecological paradigm for education. In I. Robottom (Ed.), *Environmental education: Practice and possibility* (pp. 49–68). Burwood: Deakin University.

Gough, N., & Gough, A. (2010). Environmental education. In C. Kridel (Ed.), *The SAGE encyclopedia of curriculum studies* (pp. 339–343). New York: Sage.

Grace, M., & Sharp, J. (2000). Exploring the actual and potential rhetoric-reality gaps in environmental education and their implications for pre-service teacher training. *Environmental Education Research, 6*(4), 331–345.

Gramsci, A. (1971). *Selections from the prison notebooks.* London: Lawrence and Wishart.

Greenall, A. (1980). *Environmental education for schools or how to catch environmental education.* Canberra: Curriculum Development Centre.

Greenall, A. (1987). A political history of environmental education in Australia: Snakes and ladders. In I. Robottom (Ed.), *Environmental education: Practice and possibility* (pp. 3–21). Burwood: Deakin University Press.

Gruenewald, D. A. (2003). The best of both worlds: A critical pedagogy of place. *Educational Researcher, 32*(4), 3–12.

Habermas, J. (1972). *Knowledge and human interests.* London: Heinemann.

Habermas, J. (1975). *Legitimation crisis.* Boston: Beacon Press.

Hamm, B., & Muttagi, P. K. (1998). *Sustainable development and the future of cities.* London: Intermediate Technology Publications.

Hargreaves, A. (1997). Cultures of teachers and educational change. In M. Fullan (Ed.), *The challenge of school change: A collection of articles* (pp. 47–69). Arlington Heights: IRI/Skylight Training and Publishing.

Horkheimer, M. (1982). *Critical theory.* New York: Seabury Press.

Horkheimer, M., & Adorno, T. W. (1972). *Dialectic of enlightenment.* New York: Herder and Herder.

Huckle, J. (1986). Ten red questions to ask a green teacher. *Education Links, 37,* 4–8.

Huckle, J. (1987a). *Environmental education in Australia-some perceptions and comparisons with the UK experience.* London: Australian Studies Centre.

Huckle, J. (1987b). Reflections on a trip down under. *Annual Reviews of Environmental Education, 1,* 44–46.

Huckle, J. (1991). Education for sustainability: Assessing pathways to the future. *Australian Journal of Environmental Education, 7,* 43–59.

Hungerford, H. R., & Volk, T. (1990). Changing learner behaviour through environmental education. *Journal of Environmental Education, 21*(3), 8–21.

IUCN. (1980). *World conservation strategy.* Gland: International Union for the Conservation of Nature and Natural Resources and the World Wildlife Fund.

Jickling, B. (2003). Environmental education and environmental advocacy: Revisited. *Journal of Environmental Education, 34*(2), 20–27.

Jickling, B., & Spork, H. (1998). Education for the environment: A critique. *Environmental Education Research, 4*(3), 309–328.

Kahn, R. V. (2010). *Critical pedagogy, ecoliteracy, and planetary crisis: The ecopedagogy movement.* New York: Peter Lang Publishing Inc.

Kelly, T. M. (1986). Discussing controversial issues: Four perspectives on the teacher's role. *Theory and Research in Social Education, 14*(2), 309–327.

Kemmis, S. (1991). Emancipatory action research and postmodernisms. *Curriculum Perspectives, 11*(4), 59–65.

Kemmis, S., Cole, P., & Suggett, D. (1983). *Orientations to curriculum and transition: Towards the socially-critical school*. Melbourne: Victorian Institute of Secondary Education.

Kollmuss, A., & Agyeman, J. (2002). Mind the gap: Why do people act environmentally and what are the barriers to pro-environmental behaviour? *Environmental Education Research, 8*(3), 239–260.

Kraus, S. J. (1995). Attitudes and the prediction of behavior: A meta-analysis of the empirical literature. *Journal of Personality and Social Psychology, 21*(1), 58–75.

Larri, L. (2006). *Comparative assessment: Australian Sustainable Schools Initiative pilot programme in NSW and Victoria*. Renshaw-Hitchen & Associates for the Department of the Environment and Heritage. http://pandora.nla.gov.au/pan/64516/20061018-0000/www.deh.gov.au/education/publications/comparative-assessment.html. Accessed 1 Oct 2014.

Lee, J. C.-K., & Williams, M. (2001). Researching environmental education in the school curriculum: An introduction for students and teacher researchers. *International Research in Geographical and Environmental Education, 10*(3), 218–244.

Lewin, K., & Grabbe, P. (1945). Conduct, knowledge, and acceptance of new values. *Journal of Social Issues, 1*, 53–64.

Linke, R. (1980). *Environmental education in Australia*. Sydney: Allen and Unwin.

Linke, R. (1984). Reflections on environmental education: Past development and future concepts. *Australian Journal of Environmental Education, 1*(1), 2–4.

Lucas, A. M. (1972). *Environment and environmental education: conceptual issues and curriculum implications*. PhD, Ohio State University, Ohio, USA.

Lucas, A. M. (1979). *Environment and environment education: Conceptual issues and curriculum interpretations*. Kew: Australia International Press.

Lucas, A. M. (1980). Science and environmental education: Pious hopes, self praise and disciplinary chauvinism. *Studies in Science Education, 7*, 1–21.

Luke, T. W. (2003). Critical theory and the environment. In M. Peters, C. Lankshear, & M. Olssen (Eds.), *Critical theory and the human condition. Founders and praxis* (pp. 238–250). New York: Peter Lang Publishing.

Manteaw, O. O. (2012). Education for sustainable development in Africa: The search for pedagogical logic. *International Journal of Educational Development, 32*, 376–383.

Martin, G. C. (1975). A review of objectives for environmental education. In G. C. Martin & K. Wheeler (Eds.), *Insights into environmental education* (pp. 20–32). Edinburgh: Oliver and Boyd.

Matthews, J. (2011). Hybrid pedagogies for sustainability education. *The Review of Education, Pedagogy and Cultural Studies, 33*, 260–277.

MCEETYA. (1999). *The Adelaide declaration on national goals for schooling in the twenty-first century*. Carlton: Ministerial Council on Education, Employment, Training and Youth Affairs. http://www.abs.gov.au/ausstats/abs@.nsf/Previousproducts/1301.0Feature%20Article232001?opendocument&tabname=Summary&prodno=1301.0&issue=2001&num=&view=. Accessed 1 Oct 2014.

McKenzie-Mohr, D., & Smith, W. A. (1999). *Fostering sustainable behaviour: An introduction to community-based social marketing*. Gabriola Island: New Society Publishers.

McKeown, R. (2002, July). *Education for sustainable development toolkit*, version 2. Knoxville: University of Tennessee, Energy, Environment and Resources Center. http://www.esdtoolkit.org. Accessed 30 Sept 2014.

McLaren, P. (1998). *Life in schools: An introduction to critical pedagogy in the foundations of education* (3rd ed.). New York: Longman.

Mogensen, F. (1997). Critical thinking: A central element in developing action competence in health and environmental education. *Health Education Research, 12*(4), 429–436.

Morrell, A., & O'Connor, M. (2002). Introduction. In E. O'Sullivan, A. Morrell, & M. O'Connor (Eds.), *Expanding the boundaries of transformative learning: Essays on theory and praxis* (pp. xv–xx). New York: Palgrave Macmillan.

O'Riordan, T. (1989). The challenge for environmentalism. In R. Peet & N. Thrift (Eds.), *New models in geography* (Vol. 1, pp. 77–102). London: Unwin and Hyman.

O'Sullivan, E., Morrell, A., & O'Connor, M. (Eds.). (2002). *Expanding the boundaries of transformative learning: Essays on theory and praxis.* New York: Palgrave.

OCED. (1999). *Education and learning for sustainable consumption.* Paris: Centre for Educational Research and Innovation.

Orr, D. (1999). Rethinking education. *The Ecologist, 29*(3), 232–234.

Peters, M., Lankshear, C., & Olssen, M. (2003). Introduction: Critical theory and the environment. In M. Peters, C. Lankshear, & M. Olssen (Eds.), *Critical theory and the human condition. Founders and praxis* (pp. 1–14). New York: Peter Lang Publishing.

Robertson, C. L., & Krugly-Smolska, E. (1997). Gaps between advocated practices and teaching realities in environmental education. *Environmental Education Research, 3*(3), 311–326.

Robottom, I. (1983). *The environmental education project evaluation report* (Vol. 1). Canberra: Curriculum Development Centre.

Rokeach, M. (1973). *The nature of human values.* New York: The Free Press.

Schugurensky, D. (2002). Transformative learning and transformative politics. The pedagogical dimension of participatory democracy and social action. In E. O'Sullivan, A. Morrell, & M. A. O'Connor (Eds.), *Expanding the boundaries of transformative learning. Essays on theory and praxis* (pp. 59–76). New York: Palgrave.

Scott, W., & Gough, S. (2003). *Sustainable development and learning: Framing the issues.* London: Routledge.

Scott, W., & Gough, S. (2004). *Key issues in sustainable development and learning: A critical review.* London: Routledge.

Scott, W., & Oulton, C. (1999). Environmental education: Arguing the case for multiple approaches. *Educational Studies, 25*(1), 89–97.

Sørensen, N. H. (1997). The problem of parallelism: A problem for pedagogic research and development seen from the perspective of environmental and health education. *Environmental Education Research, 3*(2), 179–187.

Spork, H. (1992). Environmental education: A mismatch between theory and practice. *Australian Journal of Environmental Education, 8*, 147–166.

Spring, J. (2004). *How educational ideologies are shaping global society: Intergovernmental organisations, NGOs, and the decline of the nation-state.* Mahwah: Lawrence Erlbaum.

Stapp, W. B., & Stapp, G. L. (1983). A summary of environmental education in Australia. *Australian Association for Environmental Education Newsletter, 12*, 4–6.

Sterling, S. (2003). *Whole systems thinking as a basis for paradigm change in education: Explorations in the context of sustainability.* PhD, University of Bath, UK.

Stevenson, R. B. (1986). Curriculum materials for United States and Australian schools: An explanation of the theory-practice gap in environmental education. In D. A. Cox & W. B. Stapp (Eds.), *International perspectives on environmental education: Issues and actions* (pp. 69–82). Ohio: North American Association for Environmental Education.

Stevenson, R. B. (1987). Schooling and environmental education: Contradictions in purpose and practice. In I. Robottom (Ed.), *Environmental education: Practice and possibility* (pp. 69–82). Burwood: Deakin University Press.

Stevenson, R. B. (2007a). Editorial. *Environmental Education Research, 13*(2), 129–138.

Stevenson, R. B. (2007b). Schooling and environmental/sustainability education: From discourses of policy and practice to discourses of professional learning. *Environmental Education Research, 13*(2), 265–285.

Swain, A. (2005). *Education as social action.* London: Palgrave MacMillan.

Thomas, G. J. (2005). Facilitation in education *for* the environment. *Australian Journal of Environmental Education, 21*, 107–116.

Tilbury, D., & Wortman, D. (2004). *Engaging people in sustainability.* Cambridge: Commission on Education and Communication, International Union for Conservation of Nature and Natural Resources.

Tilbury, D., Hamu, D., & Goldstein, W. (2002). Learning for sustainable development. In *'Education' earth year report on the world summit* (pp. 9.0–9.3). Gland: International Union for the Conservation of Nature.

Tilbury, D., Coleman, V., & Garlick, D. (2004). *A national review of environmental education and its contribution to sustainability in Australia*. Canberra: Report prepared by Macquarie University for The Department of Environment and Heritage.

UNCED. (1992, June 3–14). *Earth Summit. Agenda 21. The Rio Declaration on environment and development*. In United Nations conference on environment and development, promoting education and public awareness and training, Rio de Janeiro, Brazil. http://sustainabledevelopment.un.org/content/documents/Agenda21.pdf. Accessed 1 Oct 2014.

UNESCO. (1975). *The Belgrade Charter: A global framework for environmental education*. United Nations Educational, Scientific and Cultural Organisation. http://portal.unesco.org/education/en/files/33037/10935069533The_Belgrade_Charter.pdf/The%2BBelgrade%2BCharter.pdf. Accessed 1 Oct 2014.

UNESCO. (1978). *Intergovernmental conference on environmental education, Tbilisi (USSR), 14–16 October, 1977. Final report*. Paris: United Nations Educational, Scientific and Cultural Organisation.

UNESCO. (1997a, December 8–12). *The Declaration of Thessaloniki 1997. Environment and society –Education and public awareness for sustainability*. In Third intergovernmental conference on environmental education, Thessaloniki (Greece), United Nations Educational, Scientific and Cultural Organisation. http://unesdoc.unesco.org/images/0011/001177/117772eo.pdf. Accessed 1 Oct 2014.

UNESCO. (1997b). *Educating for a sustainable future: A transdisciplinary vision for concerted action*. United Nations Educational, Scientific and Cultural Organisation. http://www.unesco.org/education/tlsf/mods/theme_a/popups/mod01t05s01.html. Accessed 1 Oct 2014.

UNESCO. (2005a). *United Nations decade of education for sustainable development 2005–2014. Draft international implementation scheme*. United Nations Educational, Scientific and Cultural Organisation. http://unesdoc.unesco.org/images/0013/001399/139937e.pdf. Accessed 1 Oct 2014.

UNESCO. (2005b). *United Nations decade of education for sustainable development 2005–2014. International implementation scheme*. United Nations Educational, Scientific and Cultural Organisation. http://unesdoc.unesco.org/images/0014/001486/148654e.pdf. Accessed 1 Oct 2014.

UNESCO. (2007, November 26–28). *Moving forward from Ahmedabad. Environmental education in the 21st century*. In Fourth intergovernmental conference on environmental education, Ahmedabad (India). United Nations Educational, Scientific and Cultural Organisation. http://www.unevoc.net/fileadmin/user_upload/docs/AhmedabadFinalRecommendations.pdf. Accessed 1 Oct 2014.

Vongalis-Macrow, A. (2007, November 26–29). *The knowledge-doing gap: A theoretical perspective on developing eco agency through education*. Paper presented at the annual international education research conference, University of Notre Dame, Fremantle, Australia.

Walker, K. (1997). Challenging critical theory in environmental education. *Environmental Education Research, 3*(2), 155–162.

WCED. (1987). *Report of the world commission on environment and development: Our common future*. http://www.un-documents.net/wced-ocf.htm. Accessed 30 Sept 2014.

Wink, J. (2000). *Critical pedagogy: Notes from the real world* (2nd ed.). New York: Addison Wesley Longman.

WSSD. (2002, August 26–September 4). *The Johannesburg plan of implementation*. The world summit on sustainable development, Johannesburg (South Africa). http://www.un-documents.net/jburgdec.htm. Accessed 30 Sept 2014.

WWF. (2012). *Living planet report 2012. Biodiversity, biocapacity and better choices*. Gland: World Wildlife Fund for Nature International. http://d2ouvy59p0dg6k.cloudfront.net/downloads/1_lpr_2012_online_full_size_single_pages_final_120516.pdf. Accessed 30 Sept 2014.

Chapter 3
Getting to Know Giddens: Structuration as an Ontological Framework

Traditional approaches to educational research have viewed teachers as either the primary determinants of their actions in the classroom, or as subjects whose actions are mostly directed by social structural forces beyond their control. This has, in effect, compartmentalised educational research findings into two groups; those that address subjective factors and those that address objective factors. During the 1970s, Anthony Giddens expressed the potential of such objective and subjective factors to interrelate to direct human action in his theory of structuration. Giddens proposed that social practices arise from structure and agency phenomena that are not only dependent on each other, but that are also so interrelated that they actually presupposed one another. Based on this 'duality of structure and agency', structuration provides an ontological framework for social interaction that reflects a dynamic interplay between structure and agency. Research guided by such an ontological framework has the potential to expose perspectives and ideas about the interplay of structure and agency within an educational setting that may otherwise have been masked by more traditional approaches.

This chapter reviews Giddens' theory of structuration as it has been represented and reported in the literature. The principles of structuration are explained and related to the field of educational research.

3.1 An Ontological Framework

Sociologists have long sought to understand human social phenomena, that is, the forces and processes that shape societies and human action (Cohen 1989, p. 9). Traditional approaches usually focused on exploring the structural aspects of a society and human action separately in order to determine their roles and relative dominance (Archer 1982). Structuralist and functionalist views of society, for example, attributed human social phenomena to various combinations of constraining and

© Springer International Publishing Switzerland 2016
J. Edwards, *Socially-critical Environmental Education in Primary Classrooms*,
International Explorations in Outdoor and Environmental Education 1,
DOI 10.1007/978-3-319-02147-8_3

directive effects of structures or systematic circumstances—objective forces not controlled by individuals (Cohen 1989, p. 9). Alternatively, voluntarist views of society attributed social phenomena predominantly to subjective factors, interpreting human behaviours through hermeneutic and phenomenological conceptual lenses, that place individuals as the primary determinants of social phenomena (Rose 1998). These opposing views formed the basis of myriad theories and models that effectively compartmentalised and isolated aspects of human social experience (Mouzelis 2000). Considered together however, these ideas suggest that apparently dominant voluntarist or structural factors are complex and dynamic and that social phenomena may be better understood using a more holistic approach that recognises relationships between human action and social structure. The validity of such relationships, and the manner in which they may be practically modelled, have long been debated (Sawyer 2002). During the 1970s, Anthony Giddens expressed the potential interrelated influences of individual character and social structure on directing human action in his theory of structuration, hereafter referred to as structuration (Giddens 1976, 1979, 1984, 1991b; Giddens and Pierson 1998). Structuration provides an ontological framework in which "structure and agency are held to be irreducible to each other and causally efficacious, yet necessarily interdependent" (Willmott 1999, p. 5). The development of structuration as an ontological framework re-dressed the traditional priority given to epistemology, or 'knowing', in sociological research, as Giddens believed that the understanding of 'being' had been incompletely and poorly explored (Stones 2005, p. 33).

Educational research informed by the ontological focus of structuration has the potential to expose perspectives and ideas about the interplay of structure and agency within an educational setting that may otherwise be masked by more traditional epistemological-based approaches (Willmott 1999, p. 5; Yates 1997). Thus, in order to better understand the factors that contribute to the development of educational rhetoric–reality gaps it is essential to investigate the influence of, and relationships between, the subjective and objective factors which influence teachers' classroom practices. This requires an exploration of the ways in which human action, or agency, underpins teachers' practices.

3.2   Human Agency

The notion of agency is crucial to sociological research, and foremost in the quest to understand the way in which teachers' practices create rhetoric–reality gaps. Agency is a complex and multifaceted concept—"an abstraction greatly underspecified, often misused, much fetishized these days by social scientists" (Comaroff and Comaroff 1997, p. 37). The term has been complicated by a plethora of definitions, including for example, 'free will' and 'intention' (Davidson 1980), and 'power' and 'resistance' (Goddard 2000; McNay 2000). Confusion has also arisen from inconsistent use of the terms 'actor' and 'agent'. Traditionally, "actor refers to a person whose action is rule-governed" whereas agent "refers to a person engaged in the exercise of power" (Ahearn 2001, p. 113). This reflects a tradition to define agency according to an assumed relationship to social structures, and therefore, within either objective or subjective social paradigms. Objective social perspectives view human agency as responses to factors that are external to the individual, including social discourses or written and unwritten laws (Arts 2000). This objective perspective assumes that humans lack agency, and that they act as "automata", such that their actions do not reflect conscious choice (Loyal and Barnes 2001, p. 507). Subjective social perspectives attribute human agency to free will. Human agency is seen to reflect personal preferences and motivations directed by values, attitudes and moral ideals (Gynnild 2002). This subjective perspective assumes that humans may "act independently of and in opposition to, structural constraints" (Loyal and Barnes 2001, p. 507).

Attempts to attribute human agency unilaterally to either objective or subjective factors do not adequately encapsulate the breadth or complexity of social phenomena (Arts 2000). In particular, this subjective–objective dualism provides little insight into many of the critical motivations and overriding characteristics of human action, particularly those which reflect:

- opposition or compliance with structural constraints;
- support of, or resistance to, power asymmetries in society;
- conscious deliberation and motivation, or unconscious, unintended and unmotivated causes;
- outcomes that are effective or unsuccessful, and which yield expected and/or unforeseen consequences;
- the degree of prior knowledge, practice and/or mastery;
- the manner in which actions are interpreted; and
- differences between individual action and the organised or collaborative action of groups (Cohen 1989; Rose 1998).

In light of these, Ahearn (2001) proposed that human agency be provisionally defined as "the socioculturally mediated capacity to act" (p. 112). This definition is adopted here because it supports the notion of structuration by acknowledging the interaction of objective and subjective factors in both determining human action and interpreting that action. However, such factors are not necessarily discrete and definable entities. Human agency is not a simple series of isolated acts, but a "continuous flow of conduct" (Giddens 1979, p. 55) in which interactions between objective and subjective factors are complex, and in most contexts, continuous and dynamic (Archer 1982; Cohen 1989). In practice this indicates that determining the degree to which multiple objective and subjective factors interact to influence any specific human action is improbable. This task is made all the more difficult by the fact that any action is also, in part, determined by the relationship between such factors and an individual. These relationships reflect an individual's understanding of relevant objective and subjective factors. In other words, individuals are knowledgeable, and their actions reflect their unique knowledge (Turner 2003a).

Thus, in order to better understand teachers' agency it is essential to investigate teachers' knowledge of the objective and subjective factors relevant to the social context in which their actions take place—their schools and classrooms. This knowledge, or "knowledgeability" (Giddens 1984, p. 21), is central to understanding teachers' practices and the factors that contribute to the development of educational rhetoric–reality gaps.

3.3 Knowledgeability

The ability of individuals to alter their actions, regardless of prevailing objective and subjective factors, requires a degree of understanding about human action and social phenomena. Many human actions arise from deliberate and conscious decisions, whereas other actions are apparently spontaneous with little or no preparatory reasoning. This has led to the idea that human actions are influenced by three distinct forms of knowledge: unconscious, conscious and non-conscious.

Unconscious knowledge is considered to consist primarily of desires—the unconscious motivational drivers of action (Loyal 2003). Stones (2005) referred to this knowledge as consisting of general dispositions, or transposable skills, which include: values (a personal version of actual or potential reality and attitudes); worldviews (beliefs, derived from personal values, about the way the world is, or should be; Rohan 2000); and "habits of speech and gesture and methodologies for adapting this generalized knowledge to a range of particular practices in particular locations in time and space" (Stones 2005, p. 88). This knowledge forms a critical component of an individual's ability to maintain ontological security: an individual's need to have a well defined identity, assisted through the development of behavioural routines in environments with stable expectations (Giddens 1979).

Conscious knowledge, often referred to as discursive or propositional knowledge, consists of the reasons or motivations able to be expressed by individuals to justify their behaviour (Loyal 2003; Polanyi 1958). Conscious knowledge forms the basis of ideologies (sets of beliefs that contain explicit and/or implicit references to values), and both honest and false explanations of behaviour (Giddens 1979; Stake 2001). Conscious knowledge also includes the conscious decisions to act in ways that prioritise the influence of different forms of knowledge. For example, the positivist approach of science deliberately places conscious or propositional knowledge above all other forms of knowledge, whereas more naturalistic endeavours consciously acknowledge the importance of non-conscious and unconscious knowledge as essential for understanding the richness of human experience (Stake 2001).

Non-conscious knowledge is also referred to as tacit knowledge, or practical consciousness, and social knowledge (Giddens 1984; Polanyi 1966). Polanyi (1966) was the first to conceptualise tacit knowledge to explain the notion that "we can know more than we can tell" (p. 4). Stones (2005) preferred to use the term conjuncturally-specific knowledge to represent practical consciousness as "an agent's knowledge of the specific context of the action" and commented that "whilst such knowledge will be perceived, made sense of, categorized, ordered and reacted to, on the basis of the general-dispositional [unconscious knowledge], it is still analytically and causally distinguishable from these more transposable dimensions" (p. 90). Non-conscious knowledge incorporates a person's general understandings of the world and their place in it (Stake 2001). This knowledge reflects the shared social and cultural expectations of particular situations and roles, or social norms, which in turn reflects the values and value priorities of individuals or social groups (Giddens 1976) and which leads to the establishment of useful daily behavioural routines. The establishment of behavioural routines relieve people of the need to deliberately or consciously assess every aspect of every daily action and enable people to non-consciously act in ways that comply with social norms (Giddens 1976). Each person's knowledge of social norms, and therefore their routinised behaviour, is a unique reflection of their life experiences and is bounded by the social contexts in which it developed. Such contexts incorporate physical aspects, in terms of time and place, as well as less tangible perceptions of relative position within society (Leonard and Insch 2005; Loyal 2003). This indicates that social norms have the ability to systematically and powerfully affect human behaviour because they guide a person's: perception of social expectations; motivation to live up to social expectations; and attitudes towards different behaviours. In light of this, non-conscious knowledge has been viewed as the "cognitive and emotive anchor of the feelings of ontological security" (Giddens 1991a, p. 36), and is the basis upon which feelings of obligation are formed (Cialdini et al. 1991).

Torff (1999) suggested that many aspects of a teacher's classroom practice may be considered intuitive, and therefore informed by non-conscious knowledge. Understanding of the role of non-conscious knowledge is therefore a precursor for understanding teachers' practices and the development of educational rhetoric–reality gaps. Each teacher fulfils a variety of roles in their daily life, such as a tennis coach, choir member or parent, in addition to being a school teacher. Each of these roles encompasses multiple "role-sets", or the "compliment of role-relationships in which persons are involved by virtue of occupying a particular social status" (Merton 1957, p. 110). For example, at school, each teacher is not only an educator, but also an employee, a professional colleague and a carer of children. Each role incorporates common types of interactions for which behavioural routines could be established, and could contribute to the development of rhetoric–reality gaps (Stones 2005). Teachers' classroom practices therefore reflect not only their unique life experiences, but also their non-conscious knowledge of the cultural expectations of schools and education. This suggests that accounts of action provided by teachers may be structured in reference to assumed expectations or perceived social norms, reflecting non-conscious ideas rather than exposing conscious decisions for their classroom practices. In addition, even the method of acquisition of a teacher's account of their practice, such as an interview, may be predicated by non-conscious views regarding the role and expectations of educational research. In other words, personal accounts of action do not necessarily reveal the full connection between action and the knowledge that influenced it.

The relationship between knowledge and action is highly complex and dynamic, because any specific action will incorporate aspects of all three types of knowledge. In addition, any human action, irrespective of the underlying motivations, tacit understandings or deliberate planning, may create both intended and unintended consequences (Giddens 1979). The study of a rural region along the south coast of Western Australia by Curry et al. (2001) provides an excellent example of unintended consequences. A decline in agricultural profitability coincided with increasing numbers of people searching for affordable rural land in which to develop an alternative, environmentally peaceful lifestyle "removed from the excesses of capitalism and consumerism" (p. 110). As increasing numbers of people relocated to rural regions, the character of those regions changed: population density and land prices increased dramatically, and city-style business economies developed to meet population needs. As the consequences of even the smallest and simplest of actions cannot be predicted, people must continuously up-date their knowledge of human action and social phenomena in order to choose actions most likely to provide desired results.

Giddens (1979) described the process of acquiring social knowledge as a "continuous flow of conduct" to reflect the fact that individuals continuously re-interpret prior knowledge in order to refine and incorporate new ideas (p. 55). This process has been referred to as intentionality (Giddens 1979), reflexivity, and reflexive monitoring (Loyal 2003). Reflexive monitoring and subsequent behaviour modification is a conscious task which enables people to build and maintain appropriate non-conscious knowledgeregarding social norms (Giddens 1979). Figure 3.1 represents

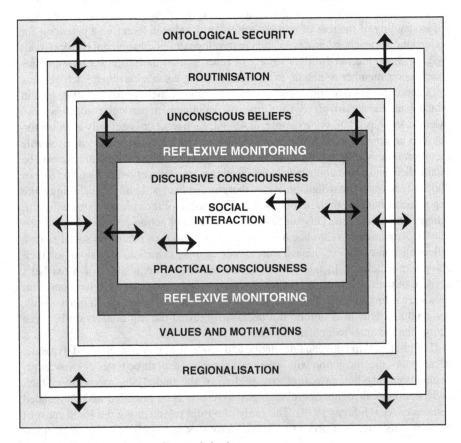

Fig. 3.1 Reflexive monitoring of human behaviour

how reflexive monitoring informs both an individual's behaviour and the context for that behaviour, and vice versa (Loyal 2003).

Giddens (1979) considered individual knowledgeability to be the vehicle through which human action could be best understood. More specifically, he perceived the identification of the boundaries of this knowledgeability, in terms of intentions, motivations and consequences, to be the primary role of any sociological study. In light of this, understanding the knowledgeability of teachers, particularly in terms of intentions and motivations, is central to developing an understanding of how teacher agency contributes to the development of educational rhetoric–reality gaps. As the practices of teachers represent a complex interrelationship between agency and the educational context in which they work, it is also essential to identify the educational structural components that work to either constrain or enable teachers' practices.

3.4 Structure

The notion of 'structure' is a complex and multifaceted aspect of social life. In sociological terms structures are abstract phenomena of pervasive social patterns or relationships, which in general, "make order out of some sets of things" (Lemert 1997, p. 127) in ways that ensure this order has a "degree of permanence" (Loyal 2003, p. 71). The term 'structure' traditionally implied that certain aspects of a society existed external to individuals, in a manner described by Lévi-Strauss for example, as underlying codes of social interactions, or "relations of presence and absence" (Loyal 2003, p. 72). However, Giddens (1984) argued that such structures were not entirely independent of individuals, nor were they able to unilaterally control human action. The ability of an individual to carry out their intended actions reflected a complex interaction between the structures of legitimation, signification- and domination—interpretation of social rules (to derive meaning and moral ideals) and the power to access and exploit required resources (degree of domination in social interactions, Giddens 1984, 1979). Thus, rules, resources and power interact to form the basic structural elements of the social interactions which constitute teachers' practices, and the classroom environments in which these practices take place.

3.5 Rules

Socio-cultural rules strongly influence human action. These are not formally defined or legally enforceable laws (although most formal laws closely reflect socio-cultural expectations). These rules incorporate informal, implied and unarticulated social expectations, or the 'social norms', that work to mediate human behaviour. In other words "rules and practices only exist in conjunction with one another" (Giddens

1979, p. 65). Rules constitute a large portion of our non-conscious knowledge and provide the foundation for contextual behavioural routines in social interaction. Although people rarely learn social practices as rules, it is through rules that people understand how to communicate and behave appropriately in different contexts (Turner 2003a). In order to understand the non-conscious factors that contribute to teachers' practices and to the development of educational rhetoric–reality gaps, it is therefore necessary to identify the socio-cultural rules embedded within their work environment; the school classroom. Such institutional rules are often associated with long-lived, well-developed practices, or routines (Loyal 2003).

Socio-cultural rules has been extensively explored by Arts (2000) and Turner (2003a) who built on Immanuel Kant's idea that there were two categories of socio-cultural rules: regulative and constitutive. Regulative rules enable people to identify the socially accepted and expected behaviours at different times, in different places, and according to the cultural character of the individuals present. Such rules reflect a society's moral expectations, and therefore enable people to legitimise their own behaviour and to judge the behaviour of others. Legitimation occurs when an individual calls upon social norms or rules in order to justify the actions of themselves or others. An example might be 'I use this text book to teach Grade 6 mathematics because *all* Grade 6 teachers use this text book'. Regulative rules inform people of their rights and obligations within different social contexts. Constitutive rules inform the way in which people interpret events in order to create signification, or derive meaning. These rules are essentially semantic—interpretative schemes of taken-for-granted understandings within different contexts. They encompass the shared understandings within a society that form the most critical elements of communication (Turner 2003a). Signification refers to the common, and usually unspoken, understandings which influence an individual's actions in particular contexts (Giddens 1979), for example, turning to face the closed door while travelling in an elevator and sitting at the back of the school bus in order to appear 'cool'. Constitutive rules underpin the way in which people organise their social interactions and make sense of actions undertaken by others (Jones et al. 2000). However, irrespective of the socio-cultural rules applicable to a context, an individual's ability to act in a preferred manner depends on their capacity to do so, and this reflects their access to resources (Turner 2003a).

3.6   Resources

Resources provide individuals with the means to interact, and are considered to be either allocative or authoritative.

Allocative resources are physical resources. The products, or raw materials, that are used in everyday life may also be used to control or direct patterns of social interaction. The unequal distribution of allocative resources contributes to unequal human relationships (Turner 2003a). On the other hand, authoritative resources are non-physical, and relate to an individual's capacity to influence, direct or organise various aspects of social interaction, such as time, space or association. Authoritative resources represent the effects of behaviours that enable individuals or groups to effectively control the pattern of interaction for a given context (Arts 2000; Taylor 2003).

Allocative and authoritative resources are complexly interrelated. An individual with greater allocative resources may enjoy elevated authority, which in turn may provide access to additional allocative resources. For example, a teacher may use their access to certain learning resources to assist their students to gain higher test scores than students taught by teachers who do not have access to similar resources. High student test scores may lead to increased communication between the teacher and the principal, which in turn, may enable the teacher to more effectively justify their need for additional resources. Human agency is thus strongly influenced by access to both material and organisational resources, which together form the structural facilities used to dominate or control social interaction (Giddens 1984). In other words, resources provide the means for obtaining power (Arts 2000; Turner 2003a). A teacher's ability to undertake a specific action, for example, to implement a socially-critical pedagogy, therefore depends in part, on their perception of, and access to, their power to do so. Thus, the identification of the effects of power relationships within teachers' work environments is an essential part of understanding the development of educational rhetoric–reality gaps.

3.7 Power

Power may be considered a "transformative capacity" because it reflects a person's ability to achieve specific outcomes from their actions (Giddens 1979, p. 88). Power is the result of the complex and dynamic interrelationship between contextually-specific rules and resources, and an individual's ability to exploit and mobilise these in order to create an asymmetric distribution of resources. Rules and resources also combine to mediate human interaction by defining social expectations for behaviour, shared meanings for communication, and appropriate sanctions for non-conformity. These in turn identify the relative power, or domination, of certain individuals in social interactions (Turner 2003a).

Human agency is therefore intrinsically related to power (Rose 1998). Traditionally, functionalist sociological approaches considered asymmetric patterns of power to be the sole determinant of human agency. Giddens (1984) however, argued that power structures are not absolute and that even the least-resourced individuals have the ability to successfully influence those who seem dominant. Knowledgeable humans have the ability to choose to use available resources in a manner that either resists or maintains a power structure, and that action leading to either of these outcomes does not imply the presence of conflict. For example, in an educational context, a teacher enjoys a certain level of power in the classroom, maintained through the continued use of well-established educational routines. A teacher may choose to relinquish some of this power by empowering students to direct their own learning, thereby altering the provision of resources and application of rules that previously defined the ways in which that teacher and those students interacted. In other words, social structures, rules and resources are "both constraining and enabling" with respect to human interaction (Giddens 1984, p. 25). Ongoing power hierarchies in social contexts reflect the complex interaction of human agency, knowledgeability, rules and resources in the establishment of regularised behavioural routines, that is, shared non-conscious knowledge. Continuing interaction, with appropriate communication and sanctions, within a specific context ensures that the presupposed power relations are maintained (Arts 2000; Giddens 1979, 1984; Loyal 2003; Taylor 2003; Turner 2003a).

3.8 Structures as Epiphenomena

Giddens (1979) believed that the socio-cultural structures that influence daily social life have no reality other than the way in which they are expressed through human action, or as they are remembered as socially expected codes of conduct. They are essentially epiphenomena: the rules and resources that reside solely within individuals as "knowledgeability in memory traces" and which are expressed only through the processes of social interaction (Stones 2005, p. 17). Figure 3.2 shows how such

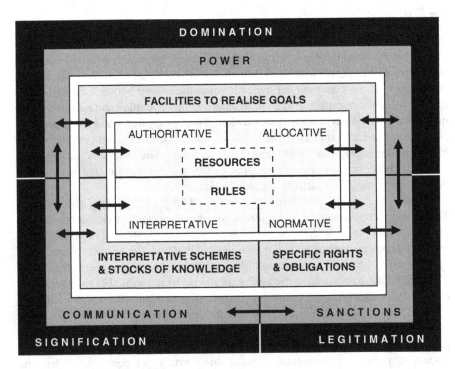

Fig. 3.2 The socio-cultural epiphenomena of domination, legitimation, and signification

epiphenomena (normative and interpretative rules, and allocative and authoritative resources) interact to form interpretative schemes (and stocks of knowledge), facilities to realise goals (access to resources) and specific rights and obligations (social norms). Such interactions, expressed and experienced as communication, sanctions and use of power, are complexly interrelated (Giddens 1984; Stones 2005):

> The facility to allocate resources is enacted in the wielding of power, and produces and reproduces social structures of domination, and moral codes (norms) help determine what can be sanctioned in human interaction, which iteratively produce structures of legitimation...thus, as human actors communicate, they draw on interpretative schemes to help make sense of interactions; at the same time those interactions reproduce and modify those interpretative schemes which are embedded in social structure as meaning or signification (Rose 2000, pp. 111–112).

According to this approach, structures not only exist just at the time and in the location in which they contribute to human action, that is, as they are "instantiated in social practices", but they also presuppose each other (Giddens 1984, p. 25). This understanding is central to Giddens' theory of structuration. Socio-cultural rules, as epiphenomena, have significant ramifications for understanding the development of the educational rhetoric–reality gaps that exist only when the social structures through which they are defined are "instantiated" in action (Giddens 1984, p. 25).

However, the notion of "structures as resources as existing only as memory traces and as instantiated in action" (Giddens 1984, p. 377) has been incompletely theorised by Giddens, resulting in a certain "lack of analytical clarity" and criticism (Stones 2005, p. 18). Archer (1996), for example, referred to material expressions of structures in the form of written laws or protocols as evidence of the tangible existence of social structures. Giddens (1979) considered these documents to be merely written representations of possibilities, in terms of the consequences or outcomes of certain actions, rather than tangible social structures. He did, however, note that it is necessary to include "certain material elements of context and capability in the notion of structure as resources" (Stones 2005, p. 18).

Archer (1995, 1996) also noted that Giddens' idea that structures exist only as they are "instantiated in action" (Giddens 1984, p. 377) failed to address a temporality in the sequence of interaction between structure and agency, and ignored the fact that structures not only exist, but are also generally longer-lived than human actions. Although her concern focused on the manner in which Giddens related structure and agency, rather than the idea that they are related, Stones (2001) suggested that this stemmed from a superficial and limited interpretation. Although Giddens did not adequately explore temporal dimensions, he did not preclude their existence— relevant social structures definitely exist, particularly at the moment they are called upon by human actions. For example, a child's ability to learn to distinguish between the taste sensations of sweet and sour will depend on the availability of food stuffs with those properties (Stones 2001).

The notion that social structures can persist over time was also supported by Cohen (1989) who stated that structures existed "as emergent properties of past practices and as the pre-existent conditions for subsequent actions" (quoted in Stones 2005, p. 63). He used the term 'position-practices' to describe sets of structures, or collections of behavioural routines, within specific contexts or institutions, such as those directing relationships between students, teachers and principals in a school. Such sets of structures exist prior to an individual entering a school, may be transformed or reproduced by that individual, and will prevail when that individual leaves (Thompson 1989). However, Archer (1995) warned that the existence of position-practices does not guarantee that every individual entering a social environment will be able to perform as expected. In other words, the practice of an individual teacher is specific to a particular place and time. That teacher may practice very differently in different places and/or at different times, even when presenting the same learning material.

It is beyond the scope of this discussion to comprehensively debate or validate the degree to which social structures exist as material resources or as knowledge. Hereafter, unless otherwise indicated, the term 'structure' refers to both material resources and knowledge, in order to most accurately represent the manner in which teachers report their perception of the interrelated nature of structures, and the influence of these on their work practices.

Teachers are knowledgeable people who continuously and reflexively monitor their actions, and who are influenced by a complex interplay of structure and human agency. Understanding the interplay of structure and human agency, particularly in terms of the role of unconscious knowledge (motivation) and non-conscious knowledge (contextual behavioural routines), is therefore essential for understanding the factors that contribute to the development of educational rhetoric–reality gaps.

3.9   Ontological Security and Routine: Where Structure and Agency Meet

Loyal and Barnes (2001) suggest that, although individuals rationalise their actions by drawing upon their knowledgeability of social structures, and reflexively monitor their actions by considering both intended and unintended consequences, neither knowledgeability, nor reflexive monitoring, explains the underlying reasons or motivation for action. Understanding human motivation has evaded decades of psychoanalytical research, and falls well outside the realms of this discussion. However, teachers' practices are undoubtedly influenced by unconscious human motivation, referred to by Giddens (1984) as "ontological security" (p. 50). Stones (2005) noted that the term was coined by Laing (1960) to describe the "inner ability of a personality to deal with threats, anxiety, ambivalence, and so on, whether in situations of the familiar routines that Giddens emphasises or in times of rapid and turbulent change" (Stones 2005, p. 24). Ontological security was considered by Giddens to be an individual's unconscious safety system; the desire to avoid negative emotions such as anxiety or guilt. In other words, people act with some reference to feelings, and in accordance with beliefs, values and attitudes (Stones 2005).

Giddens believed that ontological security is accomplished mostly through the establishment of well-practiced routines for social interaction—that is, the development of practical, non-conscious knowledge through which a significant portion of daily life is managed (Loyal 2003). The ability to follow well-established patterns of behaviour, or routinised rules, seems to diminish the importance of subjective factors in human agency, and contradicts the independent and unpredictable behaviour of 'free will' (Loyal and Barnes 2001). More than any other aspect of agency, the prevalence of such well-established routines has posed the greatest challenge for explanation and led to arguments that structural constraints direct human action. These arguments however, are countered by those who point out that routines are developed through the interaction of both objective and subjective factors of human agency (Thrift 1985; Turner 2003a).

Routines contribute predictability to daily life and enable people to interact knowingly and confidently; that is, with ontological security (Arts 2000; Turner 2003a; Vaughan 2001). The foundations of ontological security probably develop throughout childhood as an individual learns how to interact with a society's rules and resources (Kenway and Bullen 2000). In turn, the character of the behavioural routines that are created depends upon the degree to which social norms are internalised throughout an individual's socialisation process, according to, amongst other things, the perceived degree of pleasure or guilt arising from specific behaviours (Loyal and Barnes 2001). The development of social understanding and behavioural routines in this manner depends upon context, that is, the critical elements of time and space.

In order to illustrate the importance of context in social interaction, Giddens employed Heidegger's idea that "time and space represent expressions of the relations between things and events" such that "social interaction intermingles presence and absence" (quoted in Loyal 2003, p. 94). This understanding replaces the idea of 'present' with that of 'presence', such that social interaction is not characterised only by rules and resources at work, but equally by the rules and resources not being utilised. Presence, or absence, may be related to power asymmetries in any social context, indicating that routines also reflect less tangible elements such as social positioning (Loyal 2003). Understanding human agency therefore requires exploring the interrelationship of rules and resources in different contexts, and the interaction of different contexts within a social system (Gregson 1986; Thrift 1985).

The development of regionalised routines has a dual effect on social life. Not only do individuals maintain their ontological security by engaging in specific routines, their actions also facilitate the continued expectation for those routines (Giddens 1984; Mouzelis 2000; Turner 2003a; Vaughan 2001). This indicates that in addition to providing accessibility to ontological security, routines play an important role in institutionalising social structures. Thus, the practices of both the teachers and students within a classroom reflects a particular set of institutionalised routines. Any social system can be maintained for as long as individuals are willing to adhere to the routines that define and reproduce that system (Giddens 1984; Mouzelis 2000; Turner 2003a; Vaughan 2001). However, although established routines form strongly persistent aspects of social interaction, they are not fixed, and

Giddens (1984) noted that any individual may consciously decide to discard or modify any routine. In other words, reflexive monitoring may facilitate the development of useful social routines, or facilitate deliberate action contrary to established routines. Rules and resources both enable and constrain human action, but they do not determine human action (Yates 1997).

This is highlighted by Gynnild (2002) who investigated the implementation of classroom structural changes (objective factors) designed to enhance engineering students' perceptions of the worth of deep learning. Despite significant changes in the observed classes, the students' perceptions did not change. Students who initially demonstrated interest in superficial learning continued to do so, although those initially interested in deeper learning maintained their interest. A significant conclusion from this study was that efforts to alter the structural components of a learning environment did not necessarily achieve changes in student perceptions (Gynnild 2002), and that as indicated by Yates (1997), changes to the structural elements of an environment will not necessarily result in changes to human action.

Similarly, the existence of well-established routines does not necessarily ensure access to ontological security. Cassidy and Tinning (2004) illustrated the difference between intended and received messages, and the importance of the relationship between conscious and non-conscious knowledge, as individuals seek to achieve ontological security through established routines. In a study, pre-service physical education (PE) students were introduced to many images of primary school PE teachers in order to challenge their pre-conceived ideas about PE. The responses and initial teaching practices of one student were followed for the duration of the study. Both before and after viewing images of different types of physical education lessons, this student described PE teachers according to their clothing, equipment, and on-field sporting behaviour (e.g. giving directions). Despite the wide variety of images presented, this student relied upon her stereotypical views of PE teachers developed from her own experiences. She planned specific ways of being able to adhere to these stereotypical routines prior to beginning teaching.

However, the student was observed to change her planned actions in response to unexpected practices followed by her in-school mentor. For example, her mentor did not change out of sports clothing for undertaking classroom teaching. The student similarly did not change her clothing, although this created some anxiety as it was contrary to her initial ideas about what was appropriate behaviour. This demonstrated the power of an established routine to challenge ontological security, to alter individual action, and in turn, to ensure the continuance of an existing institutional practice (Cassidy and Tinning 2004).

Cassidy and Tinning (2004) also noted that the student took advantage of specific established routines within her classes. She undertook methods of student control, such as using a whistle and yelling directions, which she herself had not enjoyed at school. In other words she analysed the reality of teaching PE and took on board the shared practices of PE teachers across time and space. This demonstrated that individuals may respond to structural factors by choosing to follow a specific routine, even if required to act against personal value priorities or previously established attitudes and beliefs, and irrespective of any disruption to ontological security (Giddens 1991a).

Thus, the development of educational rhetoric–reality gaps is influenced, in part, by the interaction between an individual's feelings of ontological security and well-established teaching routines. This implies that, in order to reduce the development of such rhetoric–reality gaps, policies and programs that direct changes to teachers' practices must recognise the importance of establishing opportunities for teachers to disrupt old routines while maintaining their feelings of ontological security. In other words, reducing the development of educational rhetoric–reality gaps requires an understanding of the interaction of the important factors of teacher agency and the structural components that constitute a teacher's place of practice.

3.10   Putting It All Together: The Duality of Structure and Agency

Social theorists have traditionally viewed structure and agency as dualistic phenomena, in that individuals are either accredited with absolute freedom to act in their preferred manner, or are constrained to only those actions made possible by structural factors (Mouzelis 2000). This encouraged much sociological research to focus on determining which of the independent subjective–objective sets of factors had precedence for any specific interaction (Cassidy and Tinning 2004).

Giddens acknowledged the importance of individual choice in directing human agency. However, in order to accommodate the "patterns and predictability in action" represented by behavioural routines in daily life, he assumed that there must also be some constraints on choice due to factors which pressure individuals at the conscious or non-conscious level to choose certain actions (Loyal and Barnes 2001, p. 517). Giddens noted that structural properties may effectively place "limits upon the range of options open to an actor, or plurality of actors, in a given circumstance or type of circumstance" (Giddens 1984, pp. 176–177). Such constraints may take either of two forms: structural constraint, where action is not possible; and normative sanctions, where actions would incur punishment.

In general however, social practices arise from structure and agency phenomena that are not only dependent on each other, but are also so interrelated that Giddens assumed they actually presupposed one another (Rose 1998; Sawyer 2002). Structure and agency form a duality in which they are "mutually implicated in and constituted by the same event—social practices" (Vaughan 2001, p. 186). The notion of structure and agency as a duality "relates to the fundamentally recursive character of social life and expresses the mutual dependence of structure and agency" (Giddens 1979, p. 69). As represented in Fig. 3.2, "structural properties (signification, domination and legitimation) are constantly reproduced from social interaction (communication, power and sanctions) by means of the modalities (interpretative schemes, facilities and norms) drawn on by knowledgeable, reflexive actors" (Jones et al. 2000, p. 161). However, this notion of the duality of structure and agency has not been universally accepted. Criticism of structuration has included the concern that the notion of the duality of structure and agency precludes the possibility that individual action might occur unilaterally in response to either subjective or objective factors (e.g. Archer 1982, 1996; Sawyer 2002), and that deliberately working to reduce the distance between these introduces "crippling distortions" to any interpretation of social interaction (Mouzelis 2000). Loyal and Barnes (2001) argued that human action is best understood as a complex interplay of both chosen and unchosen factors. Irrespective of the relative importance attributed to each, no specific action can, in hindsight, be interpreted as a consequence of either choice or cause, as the action itself would be identical, and "there is only one past. Whether or not it could have been otherwise, it was not otherwise, and nothing empirical hangs on the might have been that was not" (p. 520). They proposed that the propensity for sociologists to choose either an objective or subjective perspective is merely a matter of taste, and that a more critical concern might be "what is it about theorists that makes sense of their preferences for the one or the other" (p. 520). Similarly, others have argued that continuing the debate concerning the relationship between structure and agency is futile, and that in the absence of empirical evidence that humans either exercise complete autonomy, or are totally directed by social structures, human action is more usefully understood as a range of behaviours from "extremely difficult to modify, through to those that may be modified by the most cursory intervention" (Loyal and Barnes 2001, p. 522).

Thus, teachers' classroom practices can be viewed as behavioural routines which develop from complex interactions between objective and subjective factors. Furthermore, irrespective of the degree to which subjective and objective factors influence their practices, as knowledgeable humans, teachers employ their reflexive abilities to assess socio-cultural structures and the possible implications of their actions to maintaining ontological security. In order to understand the unconscious motivations for teachers' practices and how these contribute to ontological security, it is essential to consider the human constructs of value, attitude and belief.

3.11   Unconscious Knowledge: Human Values, Attitudes and Beliefs

Reflexive monitoring is the process through which individuals determine how to act in ways that best achieve ontological security. This process requires interpretation of rules and resources, and is strongly influenced by values, attitudes and beliefs, or unconscious knowledge. An individual's unconscious knowledge will determine not only how they interpret rules and resources, but also how they will employ their agency towards ontological security (Cassidy and Tinning 2004). Understanding these complex interrelationships is necessary in order to better understand the factors that contribute to the development of educational rhetoric–reality gaps.

3.11.1 Human Values

The notion of human values is a theme central to much social science research, and yet the literature contains a long history of definitional inconsistency (Adler 1956; Campbell 1963; Rohan 2000; Smith 1969). This definitional problem reflects the difficulty in conceptualising personal constructs that are often acted upon unconsciously and that essentially describe, in part, what it is to be human (Feather 1992). Values reflect a person's perspective of their experience of their reality, referenced against a specific cultural, social and historical background (Rokeach 1973; Rohan 2000). Values are the critical components of character and personality, and enable individuals to interact uniquely to their social and physical environments (Shand 1896, 1914).

Rohan (2000) believed that this definitional confusion arose from the persistent nominalisation of the poorly understood process of valuing to the term 'value'. Humans continuously assess the relative worth (or goodness) of interactions with, and entities in, their surroundings (Festinger 1954). This cognitive process builds human unconscious knowledge—value-frameworks, or schemas, which are derived from evaluations of previous interactions, and are then used for assessing new experiences (Bargh et al. 1992; Feather 1982, 1995). Schemas are trans-situational, and enable humans to predict the outcomes of future interactions and to develop analogies for assessing unusual or unfamiliar situations (Festinger 1954). The reason humans value is evident in two value groups identified by Rokeach (1973): values that define goals that relate to personal and societal requirements (terminal), and goals for moral or competent behaviour (instrumental; Feather 1982). In other words, it is through values that individuals establish conditions for maintaining ontological security.

Raulo (2000) stated that "A person who does not know the values of his own society has no material for rational deliberation" (p. 511). Values are the standards by which individuals compare and position themselves in regard to moral and social issues, as well as religious, political and environmental ideologies. Values give individuals the means to both evaluate and rationalise their beliefs, attitudes and actions (Rokeach 1973). In other words, values are a form of unconscious knowledge and are central to the process of reflexive monitoring. The most obvious outward expression of values occurs when individuals consciously provide defensive justifications, or deliberately reframe situations, in order to disguise unacceptable differences between personal and perceived social values (Kristiansen and Hotte 1996). This has important implications for social research. It is important, for example, to consider the influence of teachers' values in the provision of descriptions of classroom practices that do not match the observed reality.

A great deal of research has been devoted to identifying a universal set of human values (e.g. Allport et al. 1960; Kohnstamm and Mervielde 1998; Morris 1956; Rokeach 1973). In several cross-cultural studies, Schwartz and Bilsky (1987a, 1990) identified many values, all of which may be grouped according to three universal human survival requirements that both individuals and groups must actively address: biological needs; social interactional demands for interpersonal coordination; and social institutional demands for group welfare and survival (Schwartz 1992, 1994, 1996). Although there is considerable debate concerning the relative influence of these groups of values, Schwartz and Bilsky (1987b) proposed that they constitute the underlying goals or concerns for all humans. Thus, identifying the values that teachers associate with the interpersonal and social demands of an educational workplace could provide valuable insights into teachers' practices, and therefore the development of educational rhetoric–reality gaps.

However, behavioural individuality is not simply a reflection of specific values, but of their relative importance or hierarchy, referred to as value priority (Kohnstamm and Mervielde 1998). In other words, when choosing how to respond to a situation, an individual will prioritise possible actions according to their value priorities. This process is not well understood, but often occurs unconsciously and involves consideration of the relative personal benefit of the probable consequences of actions (Feather 1982, 1996). Schwartz and Bilsky (1987a) suggested that this prioritising process incorporates an assessment of potential intended and unintended outcomes and consequences in relation to an individual's motivational goals, or preferences for ontological security. How an individual perceives the positive or negative aspects of an outcome will, in turn, depend on their values (Feather 1995). For example, a teacher's perception of the well-established rules or the availability of resources within their school environment may be strongly influenced by their values and value priorities. A Teacher who gives high priority to the value of tradition may be committed to maintaining the customary routines of a school, and out of respect for tradition, may accept the well-established ways in which resources are allocated. On the other hand, a teacher who gives high priority to the value of self-direction may prefer to think and act independently of the well-established traditions within a school, and may choose to create and explore new routines and new ways in which to access particular resources. In other words, a teacher's pedagogical practices are not only influenced by a school's rules and the available resources, but will also reflect that teacher's values and value priorities.

Similarly, teachers' beliefs about their ability to use available rules and resources, or ability to wield power to influence others, will influence their practices. Belief about what is possible is a strong moderating factor of behaviour (Bandura 1988). For example, a teacher may continue to instruct in an authoritative fashion, knowing that this will not encourage critical thinking in students, due to the belief that the action of one teacher cannot make enough of a difference to be worthwhile. Although a teacher's values and interpretation of potential positive and/or negative consequences are difficult to separate, it is important to consider these in order to better understand the teachers' practices that define educational rhetoric–reality gaps (Feather 1992).

Continuity in society, culture and personality suggest that human values are relatively stable, and yet the fact that social change does occur indicates that they are not permanently fixed (Rokeach 1973). As early as 1945 it was understood that value priorities are rarely altered by information alone (Lewin and Grabbe 1945). Changes are most likely to occur as a result of firsthand experiences that induce awareness by challenging an individual's ideas of what is required for ontological security. In addition, it is likely that these changes will relate specifically to the context of the experience, and probably only when that experience has been sanctioned by the individual (Rokeach 1973). Not only does this have significant implications for the goals of education *for* sustainable development (ESD), but also for recommendations to reduce educational rhetoric–reality gaps by transforming teachers' practices. If teachers are to be asked to alter their practices, they must have opportunities to experience alternatives in order to identify inconsistencies and conflicts, and to find ways in which to satisfy their needs for ontological security.

Values however, are not the sole form of unconscious knowledge that influences behaviour. Values are intimately related to attitudes, and these may therefore contribute to the development of educational rhetoric–reality gaps.

3.11.2 Attitudes and the Rhetoric–Reality Gap

Much educational research has focused on attitude, particularly in relation to teacher and student attitudes towards various subjects (e.g. Levitt 2001; Quek et al. 2007). Rokeach (1973) considered this focus to reflect: the development of survey methods believed to easily elicit attitudes from participants; a pervasive perception that attitudes strongly reflected future behaviour; and a lack of clarity regarding the difference between values and attitudes (Rokeach 1973). The latter reflects the propensity for the inconsistent use of ill-defined terminology in attitude studies. The term attitude has most often been used to refer to an individual's value judgements of both abstract situations (e.g. "I value honesty") and tangible entities (e.g. "I value this book"; Rohan 2000). The former is a judgement about a value, a type sometimes identified specifically as 'value-expressive attitudes', but which are really more simply, values (Maio and Olson 2000). The second is a judgement about an item, or action, that in itself is not a value. This type of judgement is an attitude (Rohan 2000). In other words, attitudes summarise past experiences by organising an individual's beliefs about specific situations or objects (Ajzen 1996; Rokeach 1973). Because attitudes are highly contextual they are more likely to influence certain specific behaviours than trans-situational value-based behaviours. Attitudes obviously reflect an individual's value priorities, but they may in turn bias the values an individual considers relevant to an issue or situation, and influence an individual's open-mindedness in reasoning about an issue. Individuals often employ values to justify specific attitudes (Kristiansen and Zanna 1994).

The lack of consensus about the degree to which attitudes affect behaviour reflects, in part, the complex nature of attitude formation and attitude stability (Kraus 1995). In addition, the relationship between attitudes and future behaviour is influenced by many variables, and this makes research difficult (Kraus 1995). Ajzen and Fishbein (1977) suggested that poorly designed surveys, in which there is only a weak connection between the attitudes and behaviours under investigation, have led to conclusions that attitude-behaviour correspondence is an invalid relationship.

They propose that because attitudes are highly contextual, four important elements must be considered when investigating attitude–behaviour correlation: the action, the object or target, the specific context, and the time that the action is to occur. If any one of these components does not reflect reality or the participant's perception of reality, or if any one of these components changes, there will be poor attitude–behaviour correspondence. Thus, in any research, relating attitude to behaviour depends on the reality of the situation being investigated. This, for example, means that educational rhetoric–reality gaps must be investigated as they are being created, by teachers' practices, within authentic contexts.

Although it is generally understood that attitudes are partly responsive to current contextual cues (Wood 1982), evidence suggests that both attitude and subsequent related behaviour have a common dependence on a person's prior experiences and behavioural routines. Fazio and Zanna (1981) found that attitudes created as a result of direct experience are significantly greater predictors of future behaviour than those based on indirect information. The best feedback on behaviour is from that behaviour itself. Attitudes based on prior behaviour are probably better defined and more easily evoked by future, similar situations, making them stronger (Fazio and Zanna 1981; Kraus 1995). The strength of a person's attitude associated with a specific object or action may vary on a continuum from strong to weak to non-existent. Attitude strength has been measured as time taken for individuals to react to questions concerning specific objects or situations. Well-learned, strong attitudes from direct experience provide the fastest responses because they are more readily accessed from memory (Fazio et al. 1986). The strength of a person's attitude will determine how resistant it is to change, how persistently it influences behaviour, and therefore the degree of attitude stability—probably an equally important component of attitude-behaviour predictability (Doll and Ajzen 1992; Wood 1982).

The relationship between attitude and routine suggests that altering teachers' well-established practices to reduce educational rhetoric–reality gaps requires identifying ways in which to assist teachers to identify and alter their attitudes. The strongest attitudes develop from meaningful and contextually relevant experiences. Such attitudes reflect strong object–evaluation associations which may be easily accessed from memory, and so are more likely to initiate spontaneous behaviours (Fazio et al. 1986). It is therefore essential to provide teachers with opportunities to experience new practices in ways that assist in the development of new strong attitudes that manifest as revised behavioural routines. Attitudes can be viewed as a set of beliefs that enable us to form intentions to respond (or not) to objects or situations in a particular way. Beliefs are the informational basis for our attitudes (Fishbein and Ajzen 1975). Positive attitudes develop from positive consequences and vice versa (Doll and Ajzen 1992). Attitudes act as guides rather than motivators for potential behaviour, and thus, they may predict a type of behaviour, but not that it will occur. This is an example of a rhetoric–reality gap. The identification of gaps between teachers' attitudes and their practices, in terms of the production of rhetoric–reality gaps, could provide insights into the types of interventions required to reduce these gaps.

The ability to change an attitude however, depends on first influencing the salient beliefs that contribute to the formation of that attitude. More significantly, in order to understand an individual's attitudes, it is crucial to first identify the underlying beliefs.

3.11.3   Belief

A belief is a subjective interpretation of a probable link between any two aspects of life, including objects, actions, situations, values, or concepts (Fishbein and Ajzen 1975). Rokeach (1968, 1973) described three types of beliefs: descriptive or existential beliefs which incorporate perceptions of observations and experiences; evaluative beliefs, which incorporate judgements of good or bad, and ideas of morality; and prescriptive or proscriptive beliefs which strongly influence human behaviour by enabling the consequences of actions to be judged. Despite the critical relationship between these and human attitudes, there has been little research concerning the development of these types of belief. However, there is a general understanding that beliefs form on a continuum from direct observation to inference, based on previous observations or previous inferences. The way in which a particular belief is formed will determine its strength or certainty. Fishbein and Ajzen (1975) categorised beliefs according to the manner in which they were formed: descriptive, inferential and informational.

Descriptive beliefs form as a result of evidence gained from direct observation or experience, and are generally the beliefs of greatest certainty. Because an individual's perceptions and understanding of life are influenced by their previous experiences, descriptive beliefs will, in part, reflect a continuing history of their developing understanding of the world. This suggests, for example, that a teacher may believe that a vocational/neo-classical pedagogy provides the best student learning because of their own experience as a successful student.

Inferential beliefs are formed from indirect or non-observable evidence, and are often the basis for generalisation. Many are inferred from prior descriptive beliefs, either by simple association (such as a wilted plant needs water) or as a logical progression (an emu is taller than a dingo which is taller than an echidna, so emus are taller than echidnas). Some inferential beliefs are formed from multiple interpretations, such as the inference that a student with good factual knowledge was taught by a knowledgeable teacher.

Informational beliefs are formed from reports of interpretations of situations or objects by others. The degree of belief certainty will reflect the degree to which the given interpretations are accepted. If fully accepted, such reports may be treated as direct observations leading to formation of strong beliefs (e.g. "I observed [in a government education report] that…"; Fishbein and Ajzen 1975).

The development of belief is a complex, dynamic process which begins at a very young age. Belief formation is not fully understood, and is the basis for much sociological research (e.g. Ash et al. 1993; Flavell 2000; Halstead and Taylor 2000;

Pillow and Henrichon 1996; Vinden 2002). Festinger (1954) suggested that, throughout life, individuals continuously test and up-date their beliefs. This is the basis for belief change. When an individual's belief is challenged to the point that they develop feelings of doubt about its validity, they are open to information that may clarify or alter their position. When direct observation or experience is not possible, this process involves comparison with information and feedback from others. Information from others can significantly influence belief. Consider the placebo effect (traditionally referred to in medical experiments when patients recover because they believe they are consuming medicine when they are only receiving sugar pills or placebos) and the Pygmalion effect (observed as poor performance due to convincing an able person that they are unable; O'Connor and Seymour 1995).

In light of this, teachers' practices strongly reflect, in part, their beliefs. In order to understand educational rhetoric–reality gaps, it is essential to identify the beliefs that underlie the practices that defined these gaps, including beliefs about self, beliefs about the socio-cultural rules and resources of the school setting, and beliefs about the beliefs of others.

Beliefs form the foundation of human attitudes and values, and as such, are significant for the way in which human unconscious knowledge influences perception of socio-cultural structures, reflexive monitoring, and ultimately, action. The complex interrelationship between all aspects of human knowledge and social structural elements is the basis for Giddens' theory of structuration—an ontological framework for social interaction.

3.12   Structuration as an Ontological Framework

Giddens' notion of structuration provides a theoretical perspective for understanding human phenomena by focusing on the process by which social structures are produced and reproduced over time and across space while being transformed through human interaction. This process positions structure and agency as a duality (Rose 1998; Yates 1997), and incorporates the notion of praxis.

Praxis has been described as a "somewhat ambiguous term", but is used here to represent "the use of thought to organise action to change conditions and the use of experiences in action to re-examine thought" (Turner 2003b, p. 234). This notion of praxis has been a crucial component of many traditional approaches to social research. Sztompka (1994) for example, viewed the relationship between human agency and praxis as central to social practice theories, stating that "agency and praxis are two sides of the incessant social functioning; agency actualizes in praxis, and praxis reshapes agency, which actualizes itself in changed praxis" (p. 56). Similarly, Bourdieu's (1977) ideas regarding 'habitus' highlighted the interaction between human practices and social structures as a practice theory he described as "a generative process that produces practices and representations that are conditioned by the structuring structures from which they emerge. These practices and their outcomes—whether intended or unintended—then reproduce or reconfigure the habitus" (p. 78).

These ideas highlight obvious and important similarities between established practice theory and structuration, particularly in relation to the practice of human agency, and the influence of 'habitus' or social structures. However, unlike structuration, practice theory does not provide a clear recursive relationship between structure and agency. Although human agency is viewed in relation to being shaped by social structures (an excellent framework for understanding power and inequality relations in any social system) the possibility that social structures may shape human action in ways that directly change those structures is not clearly envisaged (Ahearn 2001). Rather than focusing on human agency and social structure as discrete factors, structuration concentrates on interactions between them. This indicates that understanding social phenomena requires a holistic exploration of relationships between all social components rather than studying single aspects of society in isolation (Gregson 1987).

Key to structuration is the centrality of human actions to any social system. Human actions, as social practices, are not random, but are routinised and regionalised. Social practices are recursive activities, that is, they are enacted by knowledgeable individuals who reflect and make choices about creating and using rules and resources. Such activities occur within a framework of rules and resources, or structural properties, which form the institutionalised practices in society (Clark 1990; Giddens 1979; Jones et al. 2000). In recognition of these complex and dynamically interrelated aspects of social life, structuration takes a holistic view of social interaction, referred to as "structured-praxis" (Stones 2005, p. 19). Structured-praxis encompasses not only the social conditions that shape and facilitate human action, but also the manner in which actions are initiated, undertaken and interpreted—all aspects of human interaction in the production of social life (Cohen 1989). Structured-praxis may be considered a double hermeneutic characterised by the recursive involvement of institutions and individuals, as indicated by Giddens' notion of the duality of structure and agency, whereby the social structures created by human interaction are also influenced by social structures. Because individuals use various modalities in order to draw upon the structures of domination, legitimation and signification during social interactions, they are simultaneously contributing to the reproduction or continuance of these structures (Stones 2005; see Fig. 3.2).

As an ontological framework, structuration recognises that individuals are both social agents and social theorists with the ability to interpret and incorporate their social experience with personal knowledge and belief when deciding how to act (Giddens 1984). This is best highlighted by the role of non-conscious human knowledgeability, referred to as conjuncturally-specific knowledge, which incorporates internal structures that lie beneath that which individuals observe and present to the outside world. Conjuncturally-specific knowledge incorporates three types of understanding identified by Stones (2005) as "knowledge of the interpretative schemes, power capacities, and normative expectations and principles of the agents within context" (p. 90):

- Conjunctural knowledge of how particular positioned agents within context would interpret the actions and utterances of others (interpretative schemes);
- Conjunctural knowledge of how agents within context see their own conjuncturally-specific power capacities (power); and
- Conjunctural knowledge of how the agents within context would be likely to decide how to behave, gleaned from their perception of the fit or tension between: i) those agents' ideal normative beliefs about how they should act; and ii) how they may be pressured to act in the immediate conjuncture (Stones 2005, pp. 91–92).

It is important to note that these require an agent to interpret how others within a specific social context draw upon their 'internal knowledge', such as their 'sense of power' (see Sect. 4.1.1) and how such knowledge interacts with the broader conjuncturally-specific knowledge to influence action. In other words, the actions of one individual are influenced by the "conjuncturally-specific knowledge of networked others" who may not be directly involved in the same interactions (Stones 2005, p. 93).

The "dimensions of the duality of structure" (Giddens 1984, p. 29) interrelate through a process of structured-praxis. The three most significant relationships between the dimensions of structure, modality and social interaction were summarised by Jones et al. (2000.):

(a) systems of *signification* (structure) allow agents to *communicate* (interaction) with each other through the application of *interpretative schemes* (modality); (b) systems of *domination* (structure) enable actors to affect each other's conduct via the exercise of *power* (interaction) and the application of *facilities* such as the rules and resources (modality), although…the dialectic of control suggests that actors with apparently little 'power' can affect change; and (c) systems of *legitimation* (structure) permit the *sanctioning* (interaction) of interaction through the application of *norms* (modality) (p. 163, original italics).

Phipps (2001) added that:

Structuration processes characterise a range of social behaviours where individuals or groups of people have thought about their own and others' actions and judged them as rational; where they have learned and are using the formal and informal rules and resources for interactions, and are reaffirming them for others; and where they have experienced consequences for their actions, but are contributing all the time to a relatively stable, system-like pattern of interactions in time and space (p. 189).

These interactions reflect the full duality of structure and agency, and the notion that individuals act reflexively. Not only do rules influence a person's actions, but they are also shaped by that person's interpretation of rules and their own actions, plus their perception of the interpretations and actions of others.

An example of the duality of structure and agency in social interaction is provided by an investigation into the decline in female students and workers in Information Technology (IT). A structuration framework was employed by von Hellens et al. (2004) to analyse, through interviews, how women reinforced, transformed or were constrained by the rules and resources of the IT industry. A most significant finding was that the discourses of IT professional women were characterised by dualisms not always consistent with their lived experience (von Hellens et al. 2004). Interviews indicated that many of these dualisms related to differences in the work ethic of male versus female IT workers, and of the skills and character required for the job. Interviewees considered that, compared with male workers, female workers were more likely to be concerned with details and were better communicators, but had lesser technical skills. They also considered technological knowledge to be distinct from business knowledge. Of particular interest, interviewees indicated that their presence in the IT industry demonstrated they had overcome significant barriers, especially in relation to challenging unfavourable gender perceptions, and that they were therefore different from other females. In other words, female workers talked about their IT work in ways that reinforced the very gender differences they believed they had succeeded in overcoming (von Hellens et al. 2004). This indicates that educational rhetoric–reality gaps could reflect the teachers' practices that not only created those gaps, but in so doing, also created the structural features that inhibited the teachers' abilities to change those practices.

Giddens' theory of structuration uses the notion of a duality of structure and agency to inform an ontological framework for "the kinds of things and relations that are there to be known" in the way that humans interact and societies are constituted (Stones 2005, p. 32). Figure 3.3 shows structuration as an ontological frame-

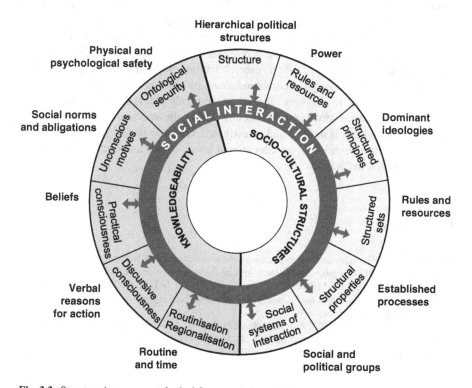

Fig. 3.3 Structuration as an ontological framework for social interaction

work for social interaction. Giddens' notion of the duality of structure and agency suggests that each of the knowledgeability and socio-cultural structural ontological elements both contribute to, and are influenced by, each other element as they are instantiated in social interaction. Figure 3.3 also shows the way in which each of the ontological elements may be expressed, or recognised, in social situations. For example: unconscious motives are generally expressed as social norms or obligations, that is, the broadly understood and socially expected ways of behaving in certain circumstances; and structured sets may be expressed as sets of rules that guide the ways in which people behave in specific circumstances.

3.13 Putting Giddens to Work

This chapter reviewed Giddens' theory of structuration which acknowledges a duality between structure and agency in the way that humans interact and societies are constituted. As such, structuration provides an ontological framework that can guide the understanding of issues represented by specific social interactions within particular social contexts, including for example, the educational issue explored in this book. Structuration provides a theoretical perspective of the critical ontological elements and interrelationships related to teachers' practices and the development of rhetoric–reality gaps in the educational environments in which education *for* sustainable development is implemented. However, in order to apply the ideals of structuration to any research issue, it is necessary to find a way in which transform these theoretical perspectives into a research methodology that can be practicably employed. In the absence of recommendations or conventions regarding how to employ structuration, researchers must adapt the ideals of this theory to effectively explore the duality of structure and agency within the specific social context in which a research issue is defined and investigated. One way in which to adapt structuration to the investigation of an educational issue, within an educational institution, is discussed in Chap. 4.

References

Adler, A. (1956). The value concept in sociology. *American Journal of Sociology, 62*, 272–279.
Ahearn, L. M. (2001). Language and agency. *Annual Reviews of Anthropology, 30*, 109–137.
Ajzen, I. (1996). The directive influence of attitudes on behavior. In P. M. Gollwitzer & J. A. Bargh (Eds.), *The psychology of action: Linking cognition and motivation to behavior* (pp. 385–403). New York: Guilford Press.
Ajzen, I., & Fishbein, M. (1977). Attitude-behavior relations: A theoretical analysis and review of empirical research. *Psychological Bulletin, 84*(5), 888–918.
Allport, G. W., Vernon, P. E., & Lindzey, G. (1960). *Study of values. Manual and test booklet* (3rd ed.). Boston: Houghton Mifflin.

Archer, M. S. (1982). Morphogenesis versus structuration: On combining structure and action. *The British Journal of Sociology, 33*(4), 455–483.

Archer, M. S. (1995). *Realist social theory: The morphogenetic approach.* Cambridge: Cambridge University Press.

Archer, M. S. (1996). Social integration and system integration: Developing the distinction. *Sociology, 30*(4), 679–700.

Arts, B. (2000). Regimes, non-state actors and the state system: A 'structurational' regime model. *European Journal of International Relations, 6*(4), 513–542.

Ash, A., Torrance, N., Lee, E., & Olson, D. (1993). The development of children's understanding of the evidence for beliefs. *Educational Psychology, 13*, 371–384.

Bandura, A. (1988). Self-regulation of motivation and action through goal system. In G. H. Bower & N. H. Frijda (Eds.), *Cognitive perspectives on emotion and motivation* (pp. 37–61). Dordrecht: Martinus Nijhoff.

Bargh, J. A., Chaiken, S., Govender, R., & Pratto, F. (1992). The generality of the automatic attitude evaluation effect. *Journal of Personality and Social Psychology, 62*, 893–912.

Bourdieu, P. (Ed.). (1977). *Outline of a theory of practice.* Cambridge: Cambridge University Press.

Campbell, D. T. (1963). Social attitudes and other acquired behavioral dispositions. In S. Koch (Ed.), *Psychology: A study of a science* (Vol. vi, pp. 94–172). New York: McGraw-Hill.

Cassidy, T., & Tinning, R. (2004). 'Slippage' is not a dirty word: Considering the usefulness of Giddens' notion of knowledgeability in understanding the possibilities for teacher education. *Teacher Education, 15*(2), 175–188.

Cialdini, R. B., Kallgren, C. A., & Reno, R. R. (1991). A focus theory of normative conduct: A theoretical refinement and reevaluation of the role of norms in human behavior. In M. P. Zanna (Ed.), *Advances in experimental social psychology* (Vol. 24, pp. 201–234). San Diego: Academic.

Clark, J. (1990). Anthony Giddens, sociology and modern social theory. In J. Clark, C. Modgil, & C. Modgil (Eds.), *Anthony Giddens, consensus and controversy* (pp. 33–45). Basingstoke: Falmer.

Cohen, I. J. (1989). *Structuration theory: Anthony Giddens and the constitution of social life* (Theoretical traditions in the social sciences). London: Macmillan.

Comaroff, J., & Comaroff, J. L. (1997). *Of revelation and revolution.* Chicago: University of Chicago Press.

Curry, G. N., Koczberski, G., & Selwood, J. (2001). Cashing out, cashing in: Rural change on the south coast of Western Australia. *Australian Geographer, 32*(1), 109–124.

Davidson, D. (1980). Agency. In D. Davidson (Ed.), *Essays on actions and events* (pp. 43–61). Oxford: Clarendon.

Doll, J., & Ajzen, I. (1992). Accessibility and stability of predictors in the theory of planned behavior. *Journal of Personality and Social Psychology, 63*(5), 754–765.

Fazio, R. H., & Zanna, M. P. (1981). Direct experience and attitude-behavior consistency. In L. Berkowitz (Ed.), *Advances in experimental social psychology* (Vol. 14, pp. 161–202). New York: Academic.

Fazio, R. H., Sanbonmatsu, D. M., Powell, M. C., & Kardes, F. R. (1986). On the automatic activation of attitudes. *Journal of Personality and Social Psychology, 50*(2), 229–238.

Feather, N. T. (1982). Human values and the prediction of action: An expectancy-valence analysis. In N. T. Feather (Ed.), *Expectations and actions: Expectancy-value models in psychology* (pp. 263–289). Hillsdale: Lawrence Erlbaum.

Feather, N. T. (1992). Values, valences, expectations and actions. *Journal of Social Issues, 48*(2), 109–124.

Feather, N. T. (1995). Values, valences, and choice: The influence of values on the perceived attractiveness and choices of alternatives. *Journal of Personality and Social Psychology, 68*(6), 1135–1151.

Feather, N. T. (1996). Values, deservingness, and attitudes toward high achievers: Research on tall poppies. In C. Seligman, J. M. Olson, & M. P. Zanna (Eds.), *The psychology of values: The Ontario symposium* (Vol. 8, pp. 215–252). Mahwah: Lawrence Erlbaum Associates.

Festinger, L. (1954). A theory of social comparison processes. *Human Relations, 7,* 117–140.

Fishbein, M., & Ajzen, I. (1975). *Belief, attitude, intention, and behavior. An introduction to theory and research.* Reading: Addison-Wesley Publishing.

Flavell, J. H. (2000). Development of children's knowledge about the mental world. *International Journal of Behavioral Development, 24*(1), 15–23.

Giddens, A. (1976). *New rules of sociological method.* New York: Hutchinson.

Giddens, A. (1979). *Central problems in social theory: Action, structure and contradiction in social analysis.* London: Macmillan.

Giddens, A. (1984). *The constitution of society.* Cambridge: Polity Press.

Giddens, A. (1991a). *Modernity and self-identity. Self and society in the late modern age.* Cambridge: Polity Press.

Giddens, A. (1991b). Structuration theory: Past, present and future. In C. G. A. Bryant & D. Jary (Eds.), *Giddens' theory of structuration: A critical appreciation* (pp. 201–221). London: Routledge.

Giddens, A., & Pierson, C. (1998). *Conversations with Anthony Giddens: Making sense of modernity.* Cambridge: Polity Press.

Goddard, V. A. (2000). *Gender, agency and change: Anthropological perspectives.* New York: Routledge.

Gregson, N. (1986). On duality and dualism: The case of structuration and time geography. *Progress in Human Geography, 10,* 184–205.

Gregson, N. (1987). Structuration theory: Some thoughts on the possibilities for empirical research. *Environment and Planning D: Society and Space, 5,* 73–91.

Gynnild, V. (2002). Agency and structure in engineering education: Perspectives on educational change in light of Anthony Giddens' structuration theory. *European Journal of Engineering Education, 27*(3), 297–303.

Halstead, J. M., & Taylor, M. J. (2000). Learning and teaching about values: A review of recent research. *Cambridge Journal of Education, 30*(2), 169–201.

Jones, O., Edwards, T., & Beckinsale, M. (2000). Technology management in a mature firm: Structuration theory and the innovation process. *Technology Analysis and Strategic Management, 12*(2), 161–177.

Kenway, J., & Bullen, E. (2000). Education in the age of uncertainty: An eagle's eye-view. *Compare, 30*(3), 265–273.

Kohnstamm, D., & Mervielde, I. (1998). Personality development. In A. Demetriou, W. Doise, & C. F. M. vanLieshout (Eds.), *Life-span developmental psychology* (pp. 399–446). Chichester: John Wiley and Sons.

Kraus, S. J. (1995). Attitudes and the prediction of behavior: A meta-analysis of the empirical literature. *Journal of Personality and Social Psychology, 21*(1), 58–75.

Kristiansen, C. M., & Hotte, A. M. (1996). Morality and the self: Implications for the when and how of value-attitude-behavior relations. In C. Seligman, J. M. Olson, & M. P. Zanna (Eds.), *The psychology of values: The Ontario symposium* (Vol. 8, pp. 77–105). Mahwah: Lawrence Erlbaum.

Kristiansen, C. M., & Zanna, M. P. (1994). The rhetorical use of values to justify social and intergroup attitudes. *Journal of Social Issues, 50*(4), 47–65.

Laing, R. D. (1960). *The divided self.* Harmondsworth: Penguin.

Lemert, C. (Ed.). (1997). *Social things.* Lanham: Rowman and Littlefield.

Leonard, N., & Insch, G. S. (2005). Tacit knowledge in academia: A proposed model and measurement scale. *Journal of Psychology, 139*(6), 495–512.

Levitt, K. E. (2001). An analysis of elementary teachers' beliefs regarding the teaching and learning of science. *Science Education, 85,* 1–22.

Lewin, K., & Grabbe, P. (1945). Conduct, knowledge, and acceptance of new values. *Journal of Social Issues, 1*, 53–64.

Loyal, S. (2003). *The sociology of Anthony Giddens*. London: Pluto.

Loyal, S., & Barnes, B. (2001). 'Agency' as a red herring in social theory. *Philosophy of the Social Sciences, 31*(4), 507–524.

Maio, G. R., & Olson, J. M. (2000). What is a 'value-expressive' attitude? In G. R. Maio & J. M. Olson (Eds.), *Why we evaluate: Functions of attitudes* (pp. 249–269). Mahwah: Lawrence Erlbaum.

McNay, L. (2000). *Gender and agency: Reconfiguring the subject in feminist and social theory*. Oxford: Blackwell.

Merton, R. K. (1957). The role-set: Problems in sociological theory. *The British Journal of Sociology, 8*(2), 106–120.

Morris, C. (1956). *Varieties of human value*. Chicago: University of Chicago Press.

Mouzelis, N. (2000). The subjectivist-objectivist divide: Against transcendence. *Sociology, 34*(4), 741–762.

O'Connor, J., & Seymour, J. (1995). *Introducing neuro-linguistic programming* (Revisedth ed.). London: Thorsons.

Phipps, A. G. (2001). Empirical application of structuration theory. *Geografiska Annaler, 83B*(4), 189–204.

Pillow, B. H., & Henrichon, A. J. (1996). There's more to the picture than meets the eye: Young children's difficulty understanding biased interpretation. *Child Development, 67*, 803–819.

Polanyi, M. (1958). *Personal knowledge: Towards a post-critical philosophy*. London: Routledge/Kegan Paul.

Polanyi, M. (1966). *The tacit dimension*. Garden City: Doubleday.

Quek, C.-L., Wong, A. F. L., Divaharan, S., Liu, W.-C., Peer, J., & Williams, M. D. (2007). Secondary school students' perceptions of teacher-student interaction and students' attitudes towards project work. *Learning Environment Research, 10*, 177–187.

Raulo, M. (2000). Moral education and development. *Journal of Social Philosophy, 31*(4), 507–518.

Rohan, M. J. (2000). A rose by any name? The values construct. *Personality and Social Psychology Review, 4*(3), 255–277.

Rokeach, M. (1968). *Beliefs, attitudes, and values*. San Francisco: Jossey-Bass.

Rokeach, M. (1973). *The nature of human values*. New York: The Free Press.

Rose, J. (1998, June 4–6). *Evaluating the contribution of structuration theory to the information systems discipline*. In Sixth European conference on information systems, Euro-Arab Management School, Aix-en-Provence, France.

Rose, J. (2000). *Information systems development as action research-soft systems methodology and structuration theory*. PhD, Manchester Metropolitan University, Manchester, UK.

Sawyer, R. K. (2002). Unresolved tensions in sociocultural theory: Analogies with contemporary sociological debates. *Culture and Psychology, 8*(3), 283–305.

Schwartz, S. H. (1992). Universals in the content and structure of values: Theoretical advances and empirical tests in 20 countries. In M. P. Zanna (Ed.), *Advances in experimental social psychology* (Vol. 25, pp. 1–65). San Diego: Academic.

Schwartz, S. H. (1994). Are there universal aspects in the structure and contents of human values? *Journal of Social Issues, 50*, 19–45.

Schwartz, S. H. (1996). Value priorities and behaviour: Applying a theory of integrated value systems. In C. Seligman, J. M. Olson, & M. P. Zanna (Eds.), *The psychology of values: The Ontario symposium* (Vol. 8, pp. 1–24). Mahwah: Lawrence Erlbaum.

Schwartz, S. H., & Bilsky, W. (1987a). Toward a psychological structure of human values. *Journal of Personality and Social Psychology, 53*, 550–562.

Schwartz, S. H., & Bilsky, W. (1987b). Toward a universal psychological structure of human values. *Journal of Personality and Social Psychology, 53*(3), 550–562.

Schwartz, S. H., & Bilsky, W. (1990). Toward a theory of the universal content and structure of values: Extensions and cross-cultural replications. *Journal of Personality and Social Psychology, 58*(5), 878–891.

Shand, A. F. (1896). Character and the emotions. *Mind, 17*, 203–226.

Shand, A. F. (1914). *The foundations of character: Being a study of the tendencies of the emotions and sentiments*. London: Macmillan.

Smith, M. B. (1969). *Social psychology and human values: Selected essays*. Chicago: Aldine.

Stake, R. E. (2001). The case study method in social inquiry. In N. K. Denzin & Y. S. Lincoln (Eds.), *The American tradition in qualitative research* (Vol. II, pp. 131–138). London: Sage.

Stones, R. (2001). Refusing the realism-structuration divide. *European Journal of Social Theory, 4*(2), 177–197.

Stones, R. (Ed.). (2005). *Structuration theory* (Traditions in social theory). London: Palgrave MacMillan.

Sztompka, P. (1994). *Agency and structure: Reorienting social theory* (Vol. 4). Langhorne: Gordon and Breach.

Taylor, V. J. (2003). Structuration revisited: A test case for an industrial archaeology methodology for far North Queensland. *Industrial Archaeology Review, XXV*(2), 129–145.

Thompson, J. B. (1989). The theory of structuration. In D. Held & J. B. Thompson (Eds.), *Social theory of modern societies: Anthony Giddens and his critics* (pp. 56–76). Cambridge: Cambridge University Press.

Thrift, N. (1985). Bear and mouse or bear and tree? Anthony Giddens' reconstruction of social theory. *Sociology, 19*(4), 609–623.

Torff, B. (1999). Tacit knowledge in teaching: Folk pedagogy and teacher education. In R. Sternberg & J. A. Horvath (Eds.), *Tacit knowledge in professional practice: Researcher and practitioner perspectives* (pp. 195–235). Mahwah: Lawrence Erlbaum.

Turner, J. H. (2003a). Structuration theory: Anthony Giddens. In *The structure of sociological theory* (7th ed., pp. 476–490). Belmont: Wadsworth.

Turner, J. H. (2003b). *The structure of sociological theory*. Belmont: Wadsworth.

Vaughan, B. (2001). Handle with care. *British Journal of Criminology, 41*, 185–200.

Vinden, P. G. (2002). Understanding minds and evidence for belief: A study of Mofu children in Cameroon. *International Journal of Behavioral Development, 26*(5), 445–452.

von Hellens, L., Nielsen, S. H., & Beekhuyzen, J. (2004). An exploration of dualisms in female perceptions of IT work. *Journal of Information Technology Education, 3*, 103–116.

Willmott, R. (1999). Structure, agency and the sociology of education: Rescuing analytical dualism. *British Journal of Sociology of Education, 20*(1), 5–21.

Wood, W. (1982). Retrieval of attitude-relevant information from memory: Effects on susceptibility to persuasion and on intrinsic motivation. *Journal of Personality and Social Psychology, 42*(5), 798–810.

Yates, J. (1997). Using Giddens' structuration theory to inform business history. *Business and Economic History, 26*(1), 159–183.

Chapter 4
Putting Giddens into Practice

Anthony Giddens' theory of structuration provides a theoretical, ontological framework for understanding social life, and as such, offers the potential to provide new perspectives of the social interactions that constitute education. However, educational researchers have been slow to embrace Giddens' ideas. This may be due to a continuing debate concerning the validity of structuration as a theoretical basis for sociological research, as well as the lack of established conventions for practicably employing structuration. This chapter reviews some critics' concerns regarding the validity of structuration. Many of these relate to the notion of the duality of structure and agency, both in terms of how well, if at all, this notion reflects real life, and whether or not it is possible to effectively assess human behaviour in terms of such a duality. Despite these concerns, Giddens' ideas are becoming incorporated into an increasing amount of social research, in fields that range from archaeology to business management. This chapter provides examples of the effective use of structuration, and highlights the fact that although Giddens' did not prescribe the knowledge to be sought, nor the methodology to be followed, in order to use structuration in practical research, the ideals of structuration can be adapted for use across a wide range of social contexts. Some of the challenges researchers face in using structuration in an educational context are discussed, and an example of how to effectively adapt the ideals of structuration to a specific research issue—the development of educational rhetoric–reality gaps—is provided.

4.1 Applying Structuration as an Ontological Framework

Structuration provides a theoretical ontological framework for taking a generic perspective on social life (Cohen 1989), but the validity of its use as a theoretical basis for sociological research is the subject of continuing and vigorous debate. Of particular concern is the limited evidence that structuration provides valid, practical

© Springer International Publishing Switzerland 2016

J. Edwards, *Socially-critical Environmental Education in Primary Classrooms*,
International Explorations in Outdoor and Environmental Education 1,
DOI 10.1007/978-3-319-02147-8_4

and ontological applicability to real social contexts (Dear and Moos 1994; Phipps 2001; Thrift 1985). Structuration is essentially a social science meta-theory, a theory that effectively encompasses others. Structuration therefore does not constrain the user to a specific research focus such as, for example, feminist or Marxist theories, nor does it attempt to yield positivist absolutes in the terms of cause and effect, or true and false (Cohen 1989; Yates 1997). The lack of a specific focus has led some (e.g. Murgatroyd 1989) to criticise structuration as lacking the critical elements of an authentic social theory. Turner (1990) agreed, noting that the lack of demonstrated normative components in the theory essentially renders structuration nothing more than a perspective of what should be, rather than what is. Thus, while structuration provides an ontological framework, it does not prescribe the knowledge to be sought, nor the methodology to be followed, in order to employ this in practical research, leaving researchers to ask "how exactly do we use the insights of structuration theory?" (Gregson 1987, p. 90). Rose (1998) added that theories are only as beneficial as their ability to guide and improve practice. Many researchers consider that structuration does not meet this criterion, and that it is no more than "an analytical scheme...a system of categories for denoting important properties of the universe", that is, merely a categorisation system for analytical comparisons (Turner 1990, p. 113).

Giddens reminded critics who wanted epistemological and methodological directions that structuration is not intended to be a method of research or a methodology, and that "the concepts of structuration theory, as with any competing theoretical perspective, should for many research purposes be regarded as sensitizing devises, nothing more" (Giddens 1984, p. 327). Giddens explained that the theory of structuration is not intended to be imported "*en bloc*" into a single empirical research (Giddens 1989, p. 294, original italics). The sensitising devises of structuration provide a mechanism for making sense of the interrelated processes that constitute social life (Giddens 1984; Turner 2003)—together these form an ontology of social life, or an "ontology of potentials":

> The structurationist ontology is addressed exclusively to the constitutive potentials of social life: the generic human capacities and fundamental conditions through which the course and outcomes of social processes and events are generated and shaped in the manifold ways in which this can occur (Cohen 1989, p. 17).

This comment reflected Giddens' idea that structures and patterns of social life exist only at the time and location that processes of human interaction occur. The importance of process prompted Sawyer (2002) to describe structuration as a "process ontology of the social world" (p. 28). Hutchins (1995) suggested that such a process ontology should be considered a socioculturalism, because culture is not formed by the collection of physical or non-physical entities, but developed from a system of processes that define the "fundamental nature of reality" (quoted in Sawyer 2002, p. 291).

Irrespective of the apparent lack of detailed information regarding how to use structuration, since its inception, its process ontology framework has been effectively employed to provide new ways of interpreting ideas from traditional fields of study. Fien (1993), for example, in an examination of how to improve education *for* the environment, identified the potential contribution of structuration "as a dialectical theory of social action for critical pedagogical practice in environmental education" (p. 13). Many researchers have used structuration to integrate qualitative and quantitative data from archival and secondary sources in order to investigate and analyse historical social phenomena. Taylor (2003), for example, employed structuration principles to interpret artefacts recovered from industrial archaeological sites in northern Queensland. Unlike traditional archaeological approaches, structuration provided insights into facets of agency within a specific historical landscape by acknowledging that structural artefacts shaped the society that simultaneously created them. Others have used a structuration ontological framework to understand the role of both structure and agency in the development of current social issues, including: power relationships within business organisations (Yates 1997); political relationships (Arts 2000); information systems technology (Jones and Karsten 2003); the interrelatedness of subjective and objective aspects of criminology (Vaughan 2001); workplace bullying as an example of specific human behaviour within a discrete context (Boucaut 2001); and the analysis of social inequalities related to geographical factors (Wilson and O'Huff 1994).

Despite the range of research problems to which the principles of structuration have been applied, a standard or preferred research approach has not been established. Structuration provides a mechanism for attaining diverse perspectives through exploration beyond a single event or action in order to incorporate the influence of both ongoing human practices and structural mechanisms (Yates 1997). In light of this, research practices must embrace the unique aspects of structuration (Stones 2005), particularly: (in)separability of structure and agency and resulting issues of temporality (Archer 1996); context; and social change (Thrift 1985). These unique features and the implications for employing a structuration ontological framework in an educational research context are discussed below.

4.1.1 (In)Separability of Structure and Agency

Social science researchers have long acknowledged the importance of both structure and agency in defining social life. Traditional research methodologies considered these to be mutually involved, but have tended to analyse them as distinct and separate influences (Archer 1996). Archer (1995) criticised Giddens' notion of the duality of structure and agency as being unable to inform social analysis because "one cannot tell where structures begin and agents end". She argued the need for a dualism where the "material and cultural conditions in which action takes place" are separated from the action itself (quoted in Stones 2005, p. 52). She indicated that such a dualistic approach was essential for exploring and explaining the relationship between structure and agency. Archer's concerns are perhaps most evident, and indeed most significant, when considering well-established routines, such as those that influence the relationship between teachers and students. In these situations the boundary between an individual's actions, internal structures and the real or perceived taken-for-granted forms of knowledge drawn upon by an individual are most blurred. It is not even clear that individuals are able to identify these boundaries, let alone a researcher (Stones 2005).

Stones (2005) provided the example of an individual drawing on structures of domination—resources of power or transformative capacity—within a particular context. A teacher within a classroom for example, has a certain sense of the power at her disposal, and the power available to others (e.g. students). These 'senses' are internal structures—virtual senses of power relations, or knowledge, drawn upon by the teacher in order to perform any action. Structuration indicates that such internal knowledge forms structures that are not only drawn upon to perform an action, but are also reflected in the manner in which the action is "instantiated" (Giddens 1984, p. 25). It is not difficult to imagine that a teacher familiar with her working environment would, over time, develop a manner of acting, or a series of routines, which reflected her internal knowledge structures and which maintained the power relations of the classroom.

Mouzelis (1991) noted that this is just one end of a continuum of the relationship between internal structures and action. He suggested that individuals are often able to describe the internal and external factors behind a specific action, but that by definition, this reflexivity required a degree of separation of subject and object (Mouzelis 2000). Similarly, any duality becomes a dualism when an individual consciously and deliberately acts to distance themselves from the rules and resources of a situation, as required for the subject–object investigative observation required within much social research. Mouzelis (2000) suggested that the relationship between any individual and the rules and resources of a context is variable, and therefore it is not possible to offer a universal statement concerning subject–object duality or dualism.

Irrespective of these arguments, simultaneously comprehending all aspects of a society is problematic (Gregson 1987). Maintaining a focus on relationships without separately characterising the interacting components and how these may change through time and across space is difficult (Rose 1998; Sawyer 2002). Stones (2001) argued that being complexly interrelated does not prohibit structure and agency from being described and understood separately. In support of this, Cassidy and Tinning (2004) suggested adopting 'methodological bracketing', an approach whereby researchers momentarily concentrate on one side of the duality in order to identify and analyse aspects of either structure or agency.

Other critics of structuration however preferred social research frameworks that embraced analytical dualism, whereby human agency and social structures are analysed separately in order to determine their relative interplay (Willmott 1999). Archer (1982) for example, presented the theory of 'morphogenesis' as a research framework that supports analytical dualism. Like structuration, morphogenesis aims to understand individuals and their social environments, that is, both the subjective and objective factors within a social system, and acknowledges a relationship between these. Developed from ideas in general systems theory (Buckley 1967), morphogenesis explores the way in which a system (a socio-cultural system) might be modified. The theoretical focus of morphogenesis is the understanding that "complex interchanges...produce change in a system's given form, structure or state" such that socio-cultural systems are essentially endless cycles of "structural conditioning/social interaction/structural elaboration" (Archer 1982, p. 458). Analytical dualism frameworks, such as morphogenesis, however, deviate significantly from structuration in their outcomes to establish the causal interactions between these factors as opposed to revealing interrelationships or processes (Sawyer 2002). The relevance of the interrelationships between factors of structure and agency to any particular research focus depends on the context in which they are revealed.

4.1.2 Context

An important aspect of structuration is the notion that social interaction is strongly dependent on context, both in time and across space. Thrift (1985) observed that despite the prominence of context, structuration itself had not been placed within a specific time or place, and that the lack of a well-developed epistemological direction presented researchers with the problem of how to move from the "level of a generalized abstract ontology—applicable to contexts of social practices at all times and places—to a particular practice situated in a particular time and place"(Stones 2005, p. 35). It is important to note that ontology-in-situ may be quite removed from ontology-in-general, and therefore it is important to identify an appropriate context for any research employing structuration as an ontological framework (Giddens 1984).

Parker (2000) argued that a structuration ontological framework is only useful for investigating the types of problems which incorporate identifiable processes able to "produce durable structures, regular patterns of interaction and development tendencies with relatively high predictability on the one hand, and volatile, unstable, randomized, quick-changing unpredictability on the other" (p. 107). In other words, the use of structuration as an ontological framework is best reserved for contexts in which the duality of structure and agency present a wide spectrum of possibilities. Despite this, Parker (2000) suggested that no single study can adequately cover every aspect of the duality of structure and agency within even the simplest context, and that therefore researchers must outline a specific investigation focus. This requires identifying both the "broader institutionalised and system-structural frame" of the research problem, and the "action horizon, as identified by the agent and/or the researcher" (Stones 2005, p. 83). For example, Thompson (1989) noted that some structures, particularly rules, take priority in different situations, and are therefore "more important than others" for resolving different research problems (Stones 2005, p. 47). This is most evident in social situations characterised by a predictable set of structures and structure priorities, or "structural identity" (Thompson 1989, p. 65). For example, the investigation of educational issues may involve teachers who work in different schools. Each school not only has a unique and distinctive set of structural characteristics, but also encompasses many structures in common with all educational institutions. Thompson (1989) suggested that the latter are not those drawn upon in the daily activities of the teachers and students but are of a "different order", existing as "a series of elements and their interrelations which together *limit* the kinds of rules which are possible and which thereby *delimit* the scope for institutional variation" (Thompson 1989, p. 66, original italics). Walsham (1998) however disagreed, stating that although the structural features of an institution may be well-established, they are maintained through the reflexive monitoring that accompanies the daily practices that define that institution, and that therefore any research must acknowledge the multilevel perspectives and influences of society, institutions and individuals.

4.1.3   Change

The way in which structuration incorporates the notion of change has been a focus for debate. The emphasis on routine in directing human agency has led to concerns that structuration is essentially a model for the process of social reproduction (Thrift 1985). The importance of routines in social life however, does not preclude social change. Even in the presence of well-established routines, individuals maintain the ability to consciously and unconsciously, and intentionally and unintentionally, act in ways that either sustain or modify routines (Yates 1997). This indicates that the modification of behavioural routines requires a change in intention and/or motivation, and possible modification to long-held value priorities, attitudes or beliefs. Many social theorists have indicated that such changes most likely occur in response to: sudden and/or unforeseen events such as death, disaster, accident or conflict; the development of new social insights or goals; and human creativity (Arts 2000; Taylor 2003; Thrift 1985). Social change therefore results from agents modifying their understanding of, or response to, previously established structures of legitimation, signification and domination (Arts 2000; Munir 2005). Irrespective of any impetus for change, the modification of an individual's routine does not predict widespread social or institutional change (Yates 1997). Giddens (1984) used the notion of episodic change to understand large-scale social change in relation to time and across space. He indicated that as every social system is composed of "recurrent social practices" (p. 66) in the form of "regularized relations of interdependence between individuals or groups" (pp. 65–66) every action will influence and change other aspects of a system in known and unknown, and intended and unintended ways. Even the most insignificant change in turn influences other actions which create change and so on (Giddens 1984). This indicates, for example, that even the presence of an observer in a classroom will undoubtedly influence and therefore change the actions of a teacher, and in turn, those changes will potentially influence the future actions of that teacher. However, Munir (2005) referred to recent changes in the photographic industry to demonstrate that social change is more complex than can be explained episodically, because different structures change at different times and within different places, such that "events in themselves are not capable of destabilising established practices" (p. 107). In other words, human reflexive monitoring continues throughout any change process, such that "actors produce sense-making schemes by either invoking existing institutional practices or by questioning them" (p. 108). This addressed Thrift's (1985) concern that structuration apparently provides little account of short-lived changes that play a significant role in any social system, in that the intended and unintended outcomes of any action or routine must satisfy a reflective appraisal prior to being repeated. Thus, if an individual's behaviour is to change, the aspects of unconscious knowledge which most strongly influence that behaviour must also change. This requires specific and authentic experiences which challenge prior understandings and established feelings of ontological security.

Some researchers have chosen to employ structuration specifically for its potential to facilitate an understanding of change processes. Structuration presents a unique perspective that as social life is constantly open to change by knowledgeable individuals, it is dynamic and not directed by universal laws. Jones et al. (2000) for example, used structuration as a framework for exploring the complex relationship of structure and agency in relation to innovation within technology companies. Their work differed from traditional studies in this field because they questioned the reasons for the appearance of new technologies, particularly in relation to what they described as "conditions under which technical change reinforces or modifies structure". This work is particularly instructive for educational research, because it explored a process through which "practices are created, developed or reinvented", and required a methodological approach with the ability to accommodate temporal dimensions (pp. 161–162).

Structuration as an ontological framework provides an excellent tool for analysing the structural and cultural interrelationships within a specific social setting (Turner, 2003, p. 488). Although Giddens' work has not typically been employed within educational research (Gynnild 2002), structuration effectively frames educational issues by highlighting the complex and dynamic interrelationships between the immediate and the broader structural and cultural influences at work within an educational environment (Rose 1998). In light of this, structuration provided an ideal ontological framework through which to investigate the research issue discussed in this book—the development of educational rhetoric–reality gaps. The process used to effectively relate the ideals of structuration to this specific research issue is outlined below.

4.2   Relating Structuration to a Research Issue

Figure 3.3 highlights the ontological elements that contribute to the duality of structure and agency in social interaction, that is, in any social interaction in any context. It is an ontology-in-general framework. However, this ontology-in-general framework does not reflect the idiosyncrasies of a specific research issue as each of the ontological elements do not contribute equally to a specific social context. Given that it is not practical, or indeed possible, for the role and effect of each ontological element to be fully understood through any one investigation (Parker 2000), it is important to identify which of the ontological elements interact to most significantly influence the research issue at hand. This begins with the development of a research specific ontology-in-situ framework. An ontology-in-situ framework indicates how the ontological elements manifest within the context in which the research issue is grounded, that is, how they relate to "particular practice[s] situated in a particular time and place" (Stones 2005, p. 35).

The development of a carefully considered ontology-in-situ framework is the most important phase of the design of any structuration-informed research. In order to generate data that will highlight the most critical relationships between the ontological elements that contribute most to a specific research issue, research activities must be guided by an ontology-in-situ framework that most closely reflects the social context in which the research issue is grounded.

The process described below provides an example of how to develop a structuration ontology-in-situ framework for a specific research issue, in this case, the development of educational rhetoric–reality gaps. Although this process is focused on an educational issue, it can be adapted to suit any research issue that requires an investigation of social interaction. The development of the most relevant and useful ontology-in-situ framework for any particular research issue requires:

- identifying the social context and the roles, or "position-practices" (Stones 2005, p. 48), through which the research issue will be investigated;
- identifying questions through which each of the ontology-in-situ elements will inform research activities to generate data;
- identifying the understanding of each of the ontology-in-situ elements to be gained from data analysis; and
- the design of the data generation techniques that most closely address the research needs as reflected by the ontology-in-situ framework.

It is important to note that the development of an ontology-in-situ framework is a dynamic process rather than a linear or fixed procedure. The more that a researcher considers, experiences and learns about a specific social context and the position-practices of individuals, or groups of individuals, within that context, the easier it is to more accurately relate the ontology-in-situ framework to the needs of the research issue. Only an ontology-in-situ framework that accurately relates the structuration ontological elements to the social context and position-practices of individuals through which a research issue is best investigated will lead to the generation of data from which an understanding of the complex and dynamic relationships between ontological elements can be gained.

4.3 Developing an Ontology-In-Situ Framework

The development of a structuration ontology-in-situ framework for understanding the development of educational rhetoric–reality gaps required careful consideration of: the ideals of structuration (Fig. 3.3); the research issue; the social context within which the research issue occurred and would be investigated (educational institutions); and which of the various position-practices of the many individuals within that context could best provide insights into the research issue.

Educational rhetoric–reality gaps reflect the complex and dynamic interactions between the knowledgeability of educators, the rhetoric of educational theory, the reality of pedagogical practice, and the socio-cultural rules and structural organisation that constitute educational institutions. The structuration ontological framework, and Giddens' notion of the duality of structure and agency, indicate that both data generation and data analysis techniques for this research issue must take into account the richness of the educational context in which rhetoric–reality gaps exist, and work not to separate variables from that context but to integrate them in order to provide a most holistic understanding (e.g. Yin 2003). However, educational institutions are extremely complex social environments. Many individuals with very different roles (e.g. curriculum advisors, principals, teachers, parents and students) interact in ways that could influence the development of educational rhetoric–reality gaps. These interactions represent sets "of structured practices which position-incumbents can and do perform", that is, routines of behaviour, or "position-practices" (Cohen 1989, p. 210). Giddens (1982) noted that when investigating such complex social situations "the most advanced form of understanding is achieved when researchers place themselves within the context being studied" because this is the most effective manner in which to "understand the viewpoints and the behaviour that characterises social actors" (p. 15). The position-practices of individual teachers in their classrooms (their usual working environment) therefore represented the most authentic context through which to identify the ways in which complexly interrelated human and structural elements of a school environment not only influenced those teachers' classroom practices, but were also shaped by their practices, and how such relationships contributed to the development of educational rhetoric–reality gaps. Thus, the investigation of the development of educational rhetoric–reality gaps necessitated consideration of "the often delicate and subtle interlacings of reflexively organized action and institutional constraint" (Giddens 1991, p. 204) in ways that acknowledged: the roles, or "position-practices" (Stones 2005, p. 48), of individual teachers in determining classroom practices; the power relations established within the schools in which these teachers practiced; and the unique or contextually specific structural components of the school environments (Giddens 1984). In order to focus on these specific position-practices, the "shared broader framework" of the school institutional environments in which "rules and resources exist and are drawn upon" could be taken as already established (Stones 2005, p. 48).

In order to begin to understand how the ontological elements interrelated in ways that contributed specifically to the development of educational rhetoric–reality gaps, it was important to define some research boundaries within the 'position-practices' of particular teachers within specific educational institutions. The implementation of the Sustainable Schools Program (SSP) provided such boundaries (see Chap. 2). The educational aim of SSP was to facilitate social transformation towards environmental sustainability through the use of a socially-critical pedagogy as effective education *for* the environment in secondary and primary schools. The implementation of SSP as a vehicle for developing socially-critical pedagogies provided opportunities to explore teachers' responses to the requirement to implement a specific practice within their usual work environment—a situation in which the prevalence of rhetoric–reality gaps has been previously reported (e.g. Bishop and Russell 1985; Fien 2001; McKeown 2002; Robertson and Krugly-Smolska 1997; Stapp and Stapp 1983; Stevenson 2007) and which appropriately bounded cases designed to investigate the educational rhetoric–reality gaps, how they were produced, and how they were experienced and understood by teachers (Stones 2005, p. 48). The development of rhetoric–reality gaps in the implementation of SSP occurred when teachers failed to employ this socially-critical pedagogy, a practice that required them to more deeply and critically assess: the role of education in shaping human–environment relationships; the intended and unintended outcomes of their pedagogical practices; and the effect of both the structural and cultural elements of their working environment on their teaching practices. In other words, insights into the development of educational rhetoric–reality gaps required an understanding of the teachers' unique experiences of the "temporal, cultural, and structural contexts" in which they practiced and in which the rhetoric–reality gaps occurred (Charmaz 2000, p. 524).

Figure 4.1 shows a structuration ontology-in-situ framework that embraces the important aspects of the research issue in terms of the social context (school classroom), the 'position-practices' of certain individuals (teachers) and appropriate research boundaries (implementation of SSP) considered here as most important for providing research insights into the development of educational rhetoric–reality gaps. Each of the ontological elements is expressed as a question. Each question informs the focus for data generation—that is, the generation of data able to provide the most valuable insights into the development of educational rhetoric–reality gaps.

As stated earlier, there is no single or definitive ontology-in-situ framework for a particular research issue. Like all social research, the knowledgeability of an individual researcher and the social context in which they are researching will not only influence the manner in which the research is framed and conducted, but will also be shaped by the research activities. Thus, an ontology-in-situ framework essentially reflects just one researcher's momentary perspective of an issue. For example, the educational rhetoric–reality gap issue could be investigated from the perspective of principals attempting to improve the pedagogical practices of the teachers in their schools, or from the perspective of the students' engagement with the pedagogical practices of their teachers. Similarly, the same issue could be investigated from the perspective of the curriculum writers and the quality of the materials provided to the schools and teachers, or in terms of parents' expectations and the socio-cultural rules of a school classroom. It is also important to consider the benefits of investigating the duality of structure and agency, represented by a specific research issue, in terms of change in both time and space. For example, an individual teacher's preferred pedagogical practices are likely to change due to such things as experience (time), stated curriculum outcomes, particular cohorts of students, or the physical and emotional aspects of a teaching environment. Importantly, an ontology-in-situ framework is a dynamic resource, and as a researcher continues to explore a particu-

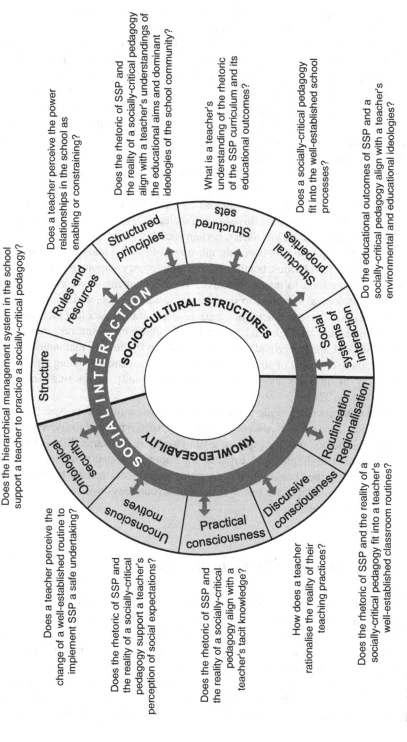

Fig. 4.1 A structuration ontology-in-situ framework: focus for data generation

lar issue the framework may be progressively up-dated, and the focus of data generation adjusted, to more effectively address the research goals.

An ontology-in-situ framework, such as that presented in Fig. 4.1, relates each ontological element to a specific research issue. This framework provides questions in order to focus data generation, but it does not specify the exact understanding or insight to be gained by answering those questions. An effective and informative ontology-in-situ framework must also reveal the reason for asking such questions, that is, it must provide a focus for data analysis. The statements given in Fig. 4.2 indicate the intended focus for data analysis for each of the questions that relate the ontological elements to the research issue. For example, Fig. 4.2 relates the ontological element 'unconscious motives' to the research issue through the question 'Does the rhetoric of SSP and the reality of a socially-critical pedagogy support a teacher's perception of social expectations?'. Figure 4.2 indicates that the understanding derived from answering this question includes 'The ways in which a teacher's practices respond to, and/or reflect, certain social expectations'.

Many of the relationships between the questions of Fig. 4.1 and the statements of Fig. 4.2 are somewhat obvious, but they form an important component of effectively employing structuration to inform research. These statements more closely align each of the ontological elements to the specific research issue, and more directly inform the design of research activities capable of generating data that incorporates the information that, through analysis, can best contribute to the research aims. Together, Figs. 4.1 and 4.2 provide an ontology-in-situ framework that indicates the way in which the concepts of structuration can inform research by being "used in a selective way in thinking about research questions [and] interpreting findings" (Giddens 1991, p. 225).

Details of the research design, including the epistemological perspective and data generation techniques used to investigate the research issue of the development of educational rhetoric–reality gaps are given elsewhere (Edwards 2011). The following discussion highlights and expands on important aspects of that research design, particularly the choice of data generation techniques, in order to provide an example of how to adapt and use a structuration ontology-in-situ framework to effectively inform research.

4.4   An Ontology-In-Situ Framework at Work

There is no preferred, or 'correct', way in which to design a research process to best address the structuration ontology-in-situ framework presented in Figs. 4.1 and 4.2. In order to investigate the development of educational rhetoric–reality gaps, a case study methodology was chosen as most appropriate for investigating the "context-dependent knowledge and experience" of teachers (Flyvbjerg 2004, p. 421). This methodology supported the ontology-in-situ framework by providing opportunities to "'close-in' on real-life situations and test views directly in relation to phenomena [rhetoric–reality gaps] as they unfold[ed] in practice" (Sørensen 1997, p. 428),

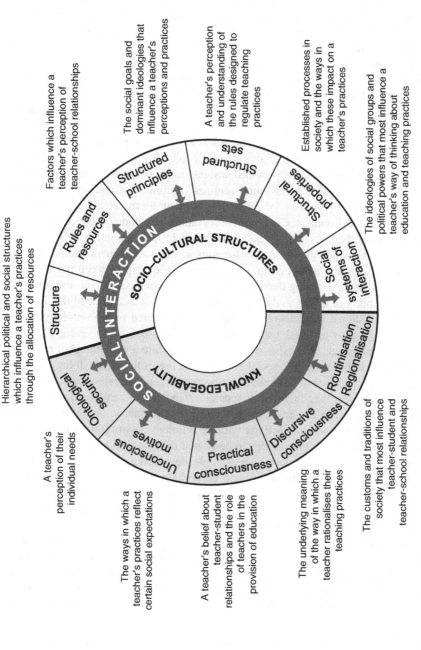

Fig. 4.2 A structuration ontology-in-situ framework: focus for data analysis

which in turn provided a rich and holistic understanding of the interrelatedness of the "temporal and spatial, historical, political, economic, cultural, social and personal" aspects of the teachers' practices and perceptions of their work (Stake 1995, p. 43). Each case study explored an authentic social context (a classroom), bounded by a "broader institutionalised and system-structural" location (a school), and the "position-practices" of an individual (a teacher) (Stones 2005, pp. 48, 83). The teachers' case studies are presented in Chap. 5.

The case study methodology also facilitated the use of a variety of data generation activities,[1] each chosen for its ability to provide insights into specific aspects of the structuration ontology-in-situ framework (Figs. 4.1 and 4.2). For example, participation in SSP professional development sessions in schools provided insights into the rules and policies (structured sets), educational aims (structured principles) and hierarchical systems (structure) that influenced the teachers' implementation of SSP. Classroom observations provided insights into critical ontological elements, including the established behavioural routines (routinisation and regionalisation), and the established processes, or constraints (structural properties), unique to each teacher's work environment. Classroom observations also contributed to understanding educational rhetoric–reality gaps through insights into the influence of not only conscious understandings (such as self-reports made in subsequent interviews), but also unconscious and non-conscious ideas (the relationship between practical and tacit knowledge) in directing the teachers' classroom practices (Silverman 2001).

The suitability of a case study methodology to the research issue of the development of educational rhetoric–reality gaps, as represented in the ontology-in-situ framework, is easily justified, particularly as the ability to employ several data generation techniques clearly assists to provide insights into a wide-range of ontological elements within a specific social context. The use of this methodology, and the incorporation of a range of participation, observation, and interview-based data generation techniques, was in no way special or specific to structuration-informed research. However, in any research activity, it is important to carefully relate the aims of each data generation activity with the ontology-in-situ framework, particularly in terms of the understandings to be gained through data analysis. This is not always a straight forward process. For example, developing a holistic understanding of an individual's knowledgeability can be problematic. The classroom observations of a teacher's practices, in conjunction with that teacher's justifications for those practices, fail to directly address the need to explore that teacher's practical consciousness (see Sect. 3.3) in order to answer the ontology-in-situ question 'Does the rhetoric of SSP and the reality of a socially-critical pedagogy align with a teacher's tacit knowledge?' (Fig. 4.1) and ultimately to understand 'A teacher's belief about teacher-student relationships and the role of teachers in the provision of education' (Fig. 4.2). The ontology-in-situ framework indicates that this understanding, as part of a teacher's knowledgeability, must have contributed to, and had been

---

[1] Data generation, including participation in professional development sessions, interviews and classroom observations were undertaken during the final school term of 2006, to coincide with the time that several schools were beginning to implement SSP.

shaped by, that teacher's classroom practices, and therefore, the development of rhetoric–reality gaps. It was therefore important to find a research technique that provided insights into each teacher's practical consciousness.

4.4.1 Investigating Teachers' Knowledgeability

Face-to-face interviews were valuable for encapsulating the teachers' perceptions of their experience of implementing SSP (Patton 1990, p. 278). Each interview provided insights into the relationship between structure and agency in the teachers' work environments, particularly in relation to ontological security, unconscious motives, practical consciousness and discursive consciousness. Interviews also highlighted aspects of both implicit and explicit rules and resources as factors influencing the teachers' classroom practices, and revealed critical aspects of social systems of interaction and structured principles directing the teachers' perceptions of their actions. However, as noted by Giddens, knowledgeability is an unfathomable mix of human expectations, motivations and perceptions (see Sect. 3.4). This means that the teachers' descriptions of, or justifications for, classroom practices provided in interviews were unlikely to completely reveal the way in which knowledge and actions were truly interrelated in the development of rhetoric–reality gaps. In order to more fully explore teachers' knowledgeability, and to provide more reliable insights into relationships between knowledgeability and structural ontological elements, hypothetical scenarios were incorporated into interviews. The use of hypothetical scenarios also addressed the observation that the teachers lacked a shared understanding, or language, for discussing educational theory in relation to pedagogy in practice.

Three short hypothetical scenarios, each of which represented specific "assertions about a possible or hypothetical reality" (Wood 2010, p. 2) of a distinctive pedagogical approach to implementing SSP were presented during interviews for discussion with teachers. Each scenario represented different understandings of knowledge, teacher–student relationships, the role of assessment, and the school community. These were modelled on the curriculum orientations work of Kemmis et al. (1983): liberal-progressive pedagogy; socially-critical pedagogy; and vocational/neo-classical pedagogy. Each hypothetical scenario, with a description of the essential characteristics of the relevant pedagogical approach, is given in Tables 4.1, 4.2 and 4.3 respectively. The use of these scenarios assisted teachers to connect "with the reality being researched" by being able to "explore circumstances" they may not have previously experienced (Wood 2010, pp. 4, 7). Hypothetical scenarios provided a neutral space for interaction, welcomed by teachers already anxious about the implementation of SSP (Van Der Heijden 2005) and allowed a degree of separation of subject and object, or methodological bracketing, which enabled teachers to consider the internal and external factors that influenced pedagogy separately (Cassidy and Tinning 2004). Morrow and Torres (2002) agreed that a degree of methodological bracketing was important, stating that "social agents and the documents of a culture must be confronted with cognitions and experiences that

Table 4.1 Hypothetical scenario: a liberal–progressive pedagogy

Hypothetical scenario (1) presented to teachers	At a school curriculum meeting, and in response to teachers' observations of improved student engagement during hands-on outdoor activities, teachers designed a project for Grade 3 students to grow native plants around the school. This project will integrate aspects of both the science (ecosystems/food webs) and SOSE (human–environment relationships) key learning areas. Teachers will organise a guest speaker from the local Indigenous community to talk about traditional use of plants, and schedule a forest walk with a parks officer to explain the roles of native plants in local environments. The class will study the local ecosystem with a view to choosing appropriate plants for inclusion in their school garden. Students will work in pairs to choose a focus for their study, and in consultation with their teacher, design learning activities. Formal assessment will require students to present research results as posters. Peer feedback will be encouraged, and final posters are to be displayed around the school for parents and visitors to view. Teachers will also note student collaboration and participation during planting activities. Garden planting might include weekend workshops with parental assistance.
Essential pedagogical characteristics	*The teacher outlines learning goals and assessment criteria. In many instances students may have a choice about how they will achieve these.*
	The teacher is responsive to student interests, concerns and prior knowledge.
	Projects are used as a method for building on important knowledge in order to gain a thorough understanding.
	SSP is being incorporated into existing science and SOSE curricula, thereby maintaining traditional subject boundaries.
	The teacher facilitates learning by organising activities and opportunities to hear from other members of society, or to visit different regions.
	The teacher's role is to arrange learning opportunities which motivate and encourage students to explore—experiences to help students make 'sense' of their world.
	Learning often occurs through experiences where students explore, problem solve, and share ideas in order to develop meaning.
	Assessment is part of the learning process, and incorporates opportunities for students to evaluate their own learning—self and/or peer assessment.
	Students may choose to undertake different aspects of the overriding project, but all work is related to the teacher's determination of the knowledge to be gained.
	Final assessment will reflect both work quality (grade) and observed development of personal skills (descriptive).
	Teacher and student negotiate the value of the knowledge and skills to be learned.
	Parents are encouraged to assist in some aspects of certain learning activities.

Table 4.2 Hypothetical scenario: a socially–critical pedagogy

Hypothetical scenario (2) presented to teachers	At a local meeting involving members of the extended school community and local residents, students canvassed ideas for establishing their role in the sustainable development of their community. Multi-age student groups explored ideas from this meeting to develop a sustainable environmental project. For example, one group chose to design and create sustainable indigenous gardens in and around local industrial/factory sites. With assistance from teachers as required, this group worked collaboratively to develop the knowledge and skills from all areas of the curriculum they needed in order to undertake this project. They negotiated with and learned from local park officers, industry owners, indigenous people, environmental groups, and local residents. Their project involved activities such as collecting indigenous seeds, propagating seedlings and designing and developing sustainable garden areas. Each student negotiated with their teacher how they might demonstrate their personal development, participation, and learning throughout the project. Teachers provided students and parents with descriptive assessments that often incorporated responses from the local community.
Essential pedagogical characteristics	*Student learning occurs through democratic participation in their community.*
	Student decision-making is an important component of learning.
	Students learn that knowledge is based on experience and reflection of self and others.
	Teachers and students are co-learners. Students work collaboratively and develop interpersonal skills alongside other learning outcomes.
	Students have opportunities to act in ways that shape their school and their society.
	Learning groups are often multi-aged and/or incorporate community members.
	Most aspects of the learning are directed by students with assistance from teachers as required.
	SSP is integrated throughout the curriculum to broaden learning experiences and to maximise opportunities for learning and acting within the community and local area.
	Teachers monitor learning to ensure students develop critical awareness of themselves and society, such that learning incorporates both understanding and action.
	Assessment requirements are negotiated between teacher and students.
	Assessment is primarily descriptive, representing personal development and community achievement. This may include evidence in the form of self, peer and community assessment.
	Students develop their understanding of self as a product of their society.

Table 4.3 Hypothetical scenario: a vocational/neo-classical pedagogy

Hypothetical scenario (3) presented to teachers	A teacher is designing lessons for a series of scheduled science classes—a program which incorporates students' interests in environmental issues while addressing essential learning requirements. This teacher has identified that, in readiness for further education, and as stipulated by government curriculum documents, Grade 6 students must understand the process of scientific inquiry. The teacher will identify reliable internet resources and appropriate books within the school library to enable students to answer the question "How do scientists work?" A parent scientist from a conservation group may visit the class to explain some aspects of their work. Students will present their answers to the class, and the teacher will compile and refine their ideas to define the scientific process. In order to demonstrate and consolidate their understanding, students will then conduct an actual investigation during science classes. Students choose from a list of environmental or sustainable development questions relating to the importance of native plants. These questions are designed by the teacher to maximise the chances of successful completion with the science laboratory resources and time available, and for their potential to provide students with an opportunity to develop essential science curriculum understandings. A template will be provided for students to use in writing a science report for assessment.
Essential pedagogical characteristics	*The teacher identifies the appropriate concepts and topics to be incorporated into any curriculum and allocates topics that combine student interests and teacher-perceived educational needs of society.*
	The overriding educational aim is to develop knowledge and skills as deemed to represent the practical requirements of society. SSP is used here to help students develop an essential understanding, namely an understanding of scientific process.
	Parental consent is required for extra-curricula activities. Parents rarely participate in day-to-day learning activities.
	Essential factual knowledge is obtained from reliable sources.
	Resources, such as books, materials or websites, are located and provided by the teacher—appropriate times are allocated during which students may access these resources.
	The role of knowledge and skills is explained in terms of an occupational perspective.
	The teacher is the authority in the classroom and uses their teaching skills to motivate students and to ensure essential knowledge is palatable to them. The teacher determines what are the most effective and efficient learning environments.
	This is a knowledge-based education. Learning is achieved through a logically structured sequence of lessons.
	The emphasis of any lesson is on producing products to be formally assessed.
	The teacher directs students to the most important knowledge and directs how students present their learning/understanding of this knowledge.
	Grades are allocated according to set criteria or standards.

allow a form of 'distanciation' from everyday reality based on explanatory accounts that elucidate the constraining and enabling effects of social structures" (p. 44).

Unannotated copies of the hypothetical scenarios were provided to teachers with the following instructions:

> Read the following three hypothetical scenarios. Each is an example of one approach that a teacher might take to implement aspects of the Sustainable Schools Program in their class-room. The topic of 'native plants' has been chosen to demonstrate these three alternative approaches and teaching styles. These are not presented in any particular order.

Each teacher was encouraged to offer comments about the scenarios as they read. When finished, and depending on the extent of their previous comments, semi-structured questions were used to prompt them to reflect more deeply upon the hypothetical scenarios, to consider aspects of the scenarios they had not commented upon, and to confirm and expand ideas as they might relate to SSP, and in turn, reflect the ideas represented in the ontology-in-situ framework.

The hypothetical scenarios enabled teachers to position themselves within educational practice theory in a manner which developed understanding through commonly shared experiences and broad understandings rather than a specific academic language. This elicited rich discussions by encouraging teachers to explore ideas most central to their own constructions and interpretations of teaching as a practice, using their own language, and unencumbered by a framework defined by specific questions (Burgess and Rudduck 1993; Merton et al. 2001). The use of the scenarios as a reflective interview process encouraged teachers to consider a broad range of ideas. Each teacher's comments were often not specific to the way in which they chose to implement SSP, but indicative of more personal ideals or unconscious knowledge (Scheurich 1997). Such unconscious knowledge was often recognised by the use of different discourses when referring to different scenarios.

Most significantly, the use of the hypothetical scenarios as part of the interview process provided valuable insights into both the conscious and non-conscious knowledgeability underlying the classroom practices and teaching ideals of the teachers, and their perspectives of the structural elements of their community, school, and classroom. Understanding these ontological aspects of each teacher's working environment was vital for exploring why teachers were implementing SSP in the ways observed and therefore central to understanding the development of educational rhetoric–reality gaps.

4.5 An Ontology-In-Situ Framework and Data Analysis

As represented in Figs. 4.1 and 4.2, a structuration ontology-in-situ framework identifies how each of the ontological elements can contribute to developing an understanding of the social interactions of a specific social context. Each ontological element, framed as a question, indicates one aspect of knowledge that could contribute to the understanding of a specific research issue. However, the notion of a duality of structure and agency represented by structuration requires exploration of the complex and dynamic relationships between the structural and hermeneutic ontological elements of a social context rather than focusing on individual elements in isolation. Giddens' did not prescribe a preferred method for the analysis of the data from structuration-informed research. The data generated from research activities informed by a well-constructed ontology-in-situ framework will reveal important relationships between individual ontological elements if the data analysis is undertaken with reference to the ideals of structuration, particularly the duality of structure and agency. Thus, irrespective of the research methods chosen, data analysis must embrace a 'structuration perspective'.

A detailed description of the data analysis process used to investigate the research issue of the development of educational rhetoric–reality gaps is given elsewhere (Edwards 2011). The following discussion uses this research issue as an example of how to embrace a structuration perspective during the analysis of data generated by techniques informed by the well-constructed ontology-in-situ framework represented in Figs. 4.1 and 4.2.

4.5.1 A 'Structuration Perspective'

In order to provide insights into the relationships between the ontological elements that most influence a specific research issue, the data generated through structuration-informed research activities requires a process of analysis that embraces a structuration perspective. A structuration perspective simply refers to the acknowledgement of the ideals of structuration, and in particular, the notion of the duality of structure and agency. Any method of data analysis can be adapted to embrace a structuration perspective, and a structuration perspective can be applied to the analysis of data that is generated from any social interaction or social context. A structuration perspective incorporates the understanding that while data generation may have addressed the statements and provided answers for the questions of an ontology-in-situ framework, it is the relationships between these aspects of the ontological elements that will contribute most to understanding the research issue.

A comparison of the relationships between the ontological elements that define the different ways in which a research issue is expressed within a social context, or which define the different position-practices of individuals within that context, can provide valuable insights into that research issue. In order for this to occur, it is essential to identify the expressions of a research issue that are most suitable for comparison. There is no right or wrong way in which to do this. In any structuration-informed research, the ways in which an issue is expressed will be intimately related to the appropriate social context, and strongly represented in the ontology-in-situ framework. For example, different expressions of the research issue of the development of educational rhetoric–reality gaps were identified through the assessment of the following aspects of the position-practices of the teachers implementing SSP:

- rhetoric—the teachers' understanding of curriculum guidance documents related to SSP and the socially-critical pedagogy embedded with them;
- reality—the manner in which the teachers were, or were not, implementing the socially-critical pedagogy of SSP; and
- the teachers' experiences of implementing SSP within their school—relationships between structure and agency in the teachers' work environments.

The data showed that the teachers held a similar understanding of the curriculum guidance documents related to SSP and the socially-critical pedagogy embedded with them (structured sets). However, the data also indicated that the position-practices of the teachers implementing SSP could be grouped according to whether or not each teacher was actually implementing a socially-critical pedagogy in conjunction with whether or not each teacher was working within a school environment that they perceived to be supportive of their efforts to implement the program (see Chap. 5). The case studies of teachers from each of these four groups provided opportunities to compare the manner in which different ontological elements contributed to the duality of structure and agency to either enable, or constrain, a teacher's practices in ways that represented either best educational practice, or an educational rhetoric–reality gap (see Chap. 6).

The data analysis technique of comparing aspects of the different expressions of a research issue is not specific to the use of structuration. In terms of the research issue of the development of rhetoric–reality gaps, such a comparison simply identified the ideas, or themes, that represented the position-practices of individual teachers and distinguished the behaviours that represented best teaching practice from the behaviours that represented rhetoric–reality gaps. Each theme could be linked to an ontological element, as highlighted by the ontology-in-situ framework that informed the research activities. Although each theme was certainly relevant to the expression of the research issue, considered in isolation these themes failed to fully explain the development of rhetoric–reality gaps. A structuration perspective required an exploration of the ways in which each of the identified themes, or ontological elements, interacted with each other. In other words, these themes had the ability to reveal the relationships between the ontological elements of the teachers' work environments that influenced the presence, or absence, of the educational rhetoric–reality gaps in the implementation of SSP. Embracing a structuration perspective provided the opportunity to develop a more holistic view of the duality of structure and agency in the development of these educational rhetoric–reality gaps (see Chap. 7).

Thus, each case study revealed a unique perspective of the dynamic and complex ontological reality of the teachers' roles as educators, and some of the most important relationships between structure and agency that the teachers indicated defined their roles, and which both enabled, and constrained, their pedagogical practices. Each of these perspectives provided valuable insights into educational rhetoric–reality gaps, and the use of a structuration perspective highlighted critical aspects of the duality of structure and agency that helped to define each of the different ways in which the research issue was expressed, through the position-practices of the teachers, within the context of implementing SSP in different schools.

As such, each case study represented just one perspective of the structuration effects of an overarching "broader institutionalised and system-structural" location, that is, an educational institution (Stones 2005, p. 83). Although the ontological elements of that broader institutional environment were not individually or specifically investigated, and were not identified by the teachers as most critical to their pedagogical choices, the ontology-in-situ framework indicated that the teachers' perceptions of those elements would certainly have influenced their practices. Applying a structuration perspective therefore required analysis of the data in order to identify how the ontological elements of that broader institutional environment, through the perspectives of the teachers, interrelated to influence the development of educational rhetoric–reality gaps. This was essential in order to identify a possible intervention point, or ontological element, through which activities and/or policies designed to reduce the development of educational rhetoric–reality gaps could be introduced into an institutional environment in which teachers work (see Chap. 8).

References

Archer, M. S. (1982). Morphogenesis versus structuration: On combining structure and action. *The British Journal of Sociology, 33*(4), 455–483.

Archer, M. S. (1995). *Realist social theory: The morphogenetic approach*. Cambridge: Cambridge University Press.

Archer, M. S. (1996). Social integration and system integration: Developing the distinction. *Sociology, 30*(4), 679–700.

Arts, B. (2000). Regimes, non-state actors and the state system: A 'structurational' regime model. *European Journal of International Relations, 6*(4), 513–542.

Bishop, J., & Russell, G. (1985). Study tours in Australia and the USA. *Review of Environmental Education Developments, 13*(2), 14–15.

Boucaut, R. (2001). Understanding workplace bullying: A practical application of Giddens' structuration theory. *International Education Journal, 2*(4), 65–73.

Buckley, W. (1967). *Sociology and modern systems theory*. Englewood Cliffs: Prentice Hall.

Burgess, R. G., & Rudduck, J. (1993). *A perspective on educational case study: A collection of papers by Lawrence Stenhouse* (Centre for Educational Development, Appraisal and Research, papers 4). Coventry: University of Warwick.

Cassidy, T., & Tinning, R. (2004). 'Slippage' is not a dirty word: Considering the usefulness of Giddens' notion of knowledgeability in understanding the possibilities for teacher education. *Teacher Education, 15*(2), 175–188.

Charmaz, K. (2000). Grounded theory: Objectivist and constructivist methods. In N. K. Denzin & Y. S. Lincoln (Eds.), *Handbook of qualitative research* (2nd ed., pp. 675–694). Thousand Oaks: Sage.

Cohen, I. J. (1989). *Structuration theory: Anthony Giddens and the constitution of social life* (Theoretical traditions in the social sciences). London: Macmillan.

Dear, M. J., & Moos, A. I. (1994). Structuration theory in urban analysis. In D. Wilson & J. O'Huff (Eds.), *Marginalized places and populations: A structurationist agenda* (pp. 3–25). Westport: Praeger Publishers.

Edwards, J. (2011). *Towards effective socially-critical environmental education: Stories from primary classrooms*. PhD thesis, RMIT University, Melbourne, Australia.

Fien, J. (1993). *Education for the environment: Critical curriculum theorising and environmental education*. Geelong: Deakin University Press.

Fien, J. (2001). *Education for sustainability: Reorientating Australian schools for a sustainable future* (Tela series). Fitzroy: Australian Conservation Foundation. http://www.acfonline.org.au/sites/default/files/resources/tela08_education_%20for_sustainability.pdf. Accessed 30 Sept 2014.

Flyvbjerg, B. (2004). Five misunderstandings about case-study research. In C. Seale, G. Gobo, J. F. Gubrium, & D. Silverman (Eds.), *Qualitative research practice* (pp. 420–434). Thousand Oaks: Sage.

Giddens, A. (1982). *Profiles and critiques in social theory*. Berkeley: University of California Press.

Giddens, A. (1984). *The constitution of society*. Cambridge: Polity Press.

Giddens, A. (1989). A reply to my critics. In D. Held & J. Thompson (Eds.), *Social theory of modern societies: Anthony Giddens and his critics* (pp. 249–301). Wiltshire: Redwood Burn.

Giddens, A. (1991). Structuration theory: Past, present and future. In C. G. A. Bryant & D. Jary (Eds.), *Giddens' theory of structuration: A critical appreciation* (pp. 201–221). London: Routledge.

Gregson, N. (1987). Structuration theory: Some thoughts on the possibilities for empirical research. *Environment and Planning D: Society and Space, 5*, 73–91.

Gynnild, V. (2002). Agency and structure in engineering education: Perspectives on educational change in light of Anthony Giddens' structuration theory. *European Journal of Engineering Education, 27*(3), 297–303.

Hutchins, E. (1995). *Cognition in the wild*. Cambridge: MIT Press.

Jones, M., & Karsten, H. (2003). *Review: Structuration theory and information systems research* (Research papers in management studies). Cambridge: Judge Institute of Management, University of Cambridge.

Jones, O., Edwards, T., & Beckinsale, M. (2000). Technology management in a mature firm: Structuration theory and the innovation process. *Technology Analysis and Strategic Management, 12*(2), 161–177.

Kemmis, S., Cole, P., & Suggett, D. (1983). *Orientations to curriculum and transition: Towards the socially-critical school*. Melbourne: Victorian Institute of Secondary Education.

McKeown, R. (2002, July). *Education for sustainable development toolkit*, version 2. Energy, Environment and Resources Center, University of Tennessee, USA. http://www.esdtoolkit.org. Accessed 30 Sept 2014.

Merton, R. K., Fiske, M., & Kendall, P. L. (2001). Purposes and criteria. In N. K. Denzin & Y. S. Lincoln (Eds.), *The American tradition in qualitative research* (pp. 328–338). London: Sage.

Morrow, R. A., & Torres, C. A. (2002). *Reading Freire and Habermas. Critical pedagogy and transformative social change*. New York: Teachers College Press.

Mouzelis, N. (1991). *Back to sociological theory: The constitution of social orders*. London: Macmillan.

Mouzelis, N. (2000). The subjectivist-objectivist divide: Against transcendence. *Sociology, 34*(4), 741–762.

Munir, K. A. (2005). The social construction of events: A study of institutional change in the photographic field. *Organization Studies, 26*(1), 93–112.

Murgatroyd, L. (1989). Only half the story: Some blinkering effects of 'malestream' sociology. In D. Held & J. B. Thompson (Eds.), *Social theories of modern societies: Anthony Giddens and his critics* (pp. 147–161). Cambridge: Cambridge University Press.

Parker, J. (2000). *Structuration theory*. Buckingham: Open University Press.

Patton, M. Q. (1990). *Qualitative evaluation and research methods*. Newbury Park: Sage.

Phipps, A. G. (2001). Empirical application of structuration theory. *Geografiska Annaler, 83B*(4), 189–204.

Robertson, C. L., & Krugly-Smolska, E. (1997). Gaps between advocated practices and teaching realities in environmental education. *Environmental Education Research, 3*(3), 311–326.

Rose, J. (1998, June 4–6). *Evaluating the contribution of structuration theory to the information systems discipline.* Sixth European Conference on Information Systems. Euro-Arab Management School, Aix-en-Provence, France.

Sawyer, R. K. (2002). Unresolved tensions in sociocultural theory: Analogies with contemporary sociological debates. *Culture and Psychology, 8*(3), 283–305.

Scheurich, J. J. (1997). *Research method in the postmodern.* London: Falmer Press.

Silverman, D. (2001). Analyzing talk and text. In N. K. Denzin & Y. S. Lincoln (Eds.), *The American tradition in qualitative research* (Vol. II, pp. 333–351). London: Sage.

Sørensen, N. H. (1997). The problem of parallelism: A problem for pedagogic research and development seen from the perspective of environmental and health education. *Environmental Education Research, 3*(2), 179–187.

Stake, R. E. (1995). *The art of case study research.* London: Sage.

Stapp, W. B., & Stapp, G. L. (1983). A summary of environmental education in Australia. *Australian Association for Environmental Education Newsletter, 12,* 4–6.

Stevenson, R. B. (2007). Schooling and environmental/sustainability education: From discourses of policy and practice to discourses of professional learning. *Environmental Education Research, 13*(2), 265–285.

Stones, R. (2001). Refusing the realism-structuration divide. *European Journal of Social Theory, 4*(2), 177–197.

Stones, R. (Ed.). (2005). *Structuration theory* (Traditions in social theory). London: Palgrave MacMillan.

Taylor, V. J. (2003). Structuration revisited: A test case for an industrial archaeology methodology for far North Queensland. *Industrial Archaeology Review, XXV*(2), 129–145.

Thompson, J. B. (1989). The theory of structuration. In D. Held & J. B. Thompson (Eds.), *Social theory of modern societies: Anthony Giddens and his critics* (pp. 56–76). Cambridge: Cambridge University Press.

Thrift, N. (1985). Bear and mouse or bear and tree? Anthony Giddens' reconstruction of social theory. *Sociology, 19*(4), 609–623.

Turner, J. H. (1990). Anthony Giddens' analysis of functionalism: A critique. In J. Clark, C. Modgil, & C. Modgil (Eds.), *Anthony Giddens, consensus and controversy* (pp. 103–114). London: Falmer.

Turner, J. H. (2003). Structuration theory: Anthony Giddens. In *The structure of sociological theory* (7th ed., pp. 476–490). Belmont: Wadsworth.

Van Der Heijden, K. (2005). *Scenarios: The art of strategic conversation* (2nd ed.). Chichester: John Wiley and Sons.

Vaughan, B. (2001). Handle with care. *British Journal of Criminology, 41,* 185–200.

Walsham, G. (1998). IT and changing professional identity: Micro-studies and macro-theory. *Journal of the American Society for Information Science, 49,* 1081–1089.

Willmott, R. (1999). Structure, agency and the sociology of education: Rescuing analytical dualism. *British Journal of Sociology of Education, 20*(1), 5–21.

Wilson, D., & O'Huff, J. (Eds.). (1994). *Marginalized places and populations: A structurationist agenda.* Westport: Praeger Publishers.

Wood, M. (2010, April 21–22). Dump the questionnaires and make it up: The value of fictional 'data' in management research. In *Fourth European Conference on Research Methods in Business and Management*, Paris, France.

Yates, J. (1997). Using Giddens' structuration theory to inform business history. *Business and Economic History, 26*(1), 159–183.

Yin, R. K. (2003). *Case study research: Design and methods* (3rd ed.). Thousand Oaks: Sage.

Chapter 5
The State of Play

The Sustainable Schools Program aims to address the goals of Education *for* Sustainable Development by establishing schools as working models of sustainable communities. In order to effectively achieve this, a school is expected to make significant changes to their operational and management procedures, and teachers and students must develop a learning environment in which they interact through a socially-critical pedagogy. The stories in this chapter document the ideas and practices of ten teachers and two principals from six Australian primary schools committed to the Sustainable Schools Program.[1] These stories highlight the experiences and perspectives of the teachers who were required to implement the socially-critical pedagogy advocated by the program. Together, these stories paint a picture of the state of the implementation of this program, and provide insights into the three main factors of educational practice which underpin the development of the educational rhetoric–reality gaps that thwart efforts to establish effective Education *for* Sustainable Development:

- Rhetoric—the teachers' understanding of the curriculum guidance documents related to the SSP and the socially-critical pedagogy embedded with them;
- Reality—the manner in which the teachers were implementing the socially-critical pedagogy of the Sustainable Schools Program; and
- Teachers' experiences of implementing the Sustainable Schools Program within their school—the relationships between structure and agency in the teachers' work environments.

---

[1] These Victorian-registered teachers and principals worked in schools identified by The Gould League and the Centre for Education and Research in Environmental Strategies as committed to implementing the Sustainable Schools Program. They represented the small minority of the educators in those schools who were willing to discuss the implementation of the program. Most of the teachers and principals of these schools were unwilling to discuss the program, citing issues of anxiety and conflict that had arisen from the need to change well-established practices.

© Springer International Publishing Switzerland 2016
J. Edwards, *Socially-critical Environmental Education in Primary Classrooms*,
International Explorations in Outdoor and Environmental Education 1,
DOI 10.1007/978-3-319-02147-8_5

5.1   Stories from an Ontology-in-situ Framework

The stories (case studies as described in Chap. 4), presented in this chapter represent the position-practices of individual teachers who were required to implement a socially-critical pedagogy as part of implementing the Sustainable Schools Program (SSP) within an authentic educational context—the schools and communities in which they worked. The insights that constitute each story reflect the questions and statements of the ontology-in-situ framework (Figs. 4.1 and 4.2) that was designed to guide the investigation of the development of educational rhetoric–reality gaps in the implementation of SSP.

The teachers nominated a teaching/learning session that they believed best portrayed their approach to implementing SSP. Each teacher's practice was assessed, according to nine criteria, to determine the degree to which it approached a socially-critical pedagogy. These criteria, identified from the descriptions of the vocational/ neo-classical (teacher-centred), liberal-progressive, and socially-critical (student-centred) pedagogies provided by Allen (2004), Huba and Freed (2000), and Kemmis et al. (1983), represent different aspects of a learning environment, particularly the roles and interrelationships of teachers and students. The degree to which each teacher's practice conformed to each of these criteria, on a scale from 0 (not at all) to 1 (closely) was then used to construct 'state of play' diagrams for the teachers at each school. In addition, the status and role of SSP within each school is described (SSP implementation), followed by a description of each teacher's understanding of the rhetoric or goals of SSP (understanding SSP) and the reality of their pedagogical implementation of these goals (SSP in practice). Each teacher's perception of different types of classroom practices is provided (understanding pedagogy), and the additional resources they identified as being essential for implementing a socially-critical pedagogy are indicated (impediments to socially-critical pedagogy).

All schools and individuals are referred to by pseudonym to maintain their right to confidentiality and anonymity.

5.2   East Valley Primary School

East Valley Primary School was a co-educational government school which catered for over 600 students. It was located approximately 60 km from Melbourne, in a semi-rural region that was undergoing significant changes due to rapidly expanding residential zones and an associated population increase. The majority of the 13,000 people living in East Valley were technicians and trades workers, most of whom commuted 10 km to work in a major industrial and manufacturing centre. In order to accommodate the increasing enrolments at East Valley Primary School, each week students attended classes in training rooms within the grounds of the neighbouring East Valley Nature Park (EVNP). In addition to EVNP, the school was located within walking distance of a range of urban, rural and natural features which included natural waterways and native bush land. Coastal landscapes and an inland state park were only a short drive from the school.

The school's overarching educational policy was to ensure that all of the students felt supported and therefore capable of succeeding at whatever they most desired to do. Learning activities were designed to build upon the students' curiosity about the world, and to foster a caring attitude towards others and the environment.

Four teachers, Anita, Julia, Karen and Robyn, shared their experiences of working to establish SSP at East Valley Primary School and to incorporate SSP ideals into their classroom practices.

Implementing SSP   East Valley Primary School was 10 months into their first year of implementing SSP. Anita, Julia, Karen and Robyn each agreed that the implementation of SSP had caused significant disruption to the school, particularly in terms of the relationships between individual teachers, and between the teachers and their principal. The implementation of SSP was directed by the principal, who felt unable to formally discuss any aspect of the implementation of the program at the school. Julia stated that SSP was being used as a vehicle for change, because the principal "wanted to point us [the teachers] into a direction" because "we were floundering". That direction, according to Anita, incorporated widespread pedagogical change through the establishment of "inquiry-based learning". Anita noted that teaching and learning at East Valley Primary School had been "very teacher-driven" and "very content-based" with little consideration of the students' interests or learning styles. In an effort to change this, and in order to better comply with the requirements of state curriculum guidelines, the principal expected the teachers to begin to incorporate such things as thinking skills, and interpersonal and intrapersonal dimensions of learning into their classroom practices. The principal hoped that the implementation of SSP would provide opportunities for the teachers to engage with, and therefore own, the change process. Anita commented that all of the teachers who were willing to attempt pedagogical change were strongly supported by the principal, but that this had created a divided workforce. She described an unpleasant work environment in which many of her colleagues were strongly resistant to change and openly dismissive of teaching and learning activities that differed from their own well-established practices. Robyn found implementing SSP highly stressful due to the lack of peer support. She valued the strong support of her principal but noted that she was unable to engage with the other early years teachers who had refused to embrace the educational change process facilitated by SSP, stating that her working environment was "very much, very, segregated. We're doing our own thing. Someone else is doing theirs…[there is] not much talk about what each of us is doing". In such an environment, Robyn did not believe that SSP could be sustained. Only those teachers most supportive of the SSP ideals participated in the professional development sessions held at both the Centre for Education and Research in Environmental Strategies (CERES) and within the school.

5.2.1   Anita

Anita had been a qualified teacher for just 3 years. She had spent this time as a generalist Prep (Preparatory year prior to Grade 1) teacher at East Valley Primary School. Anita was sharing a double classroom with Robyn.

Understanding SSP Anita stated that her participation in SSP at East Valley Primary School was "to be blunt, [be]cause we're doing inquiry-based learning right through the school". She believed that because the principal wished to better develop the students' understanding that "what we've got on Earth is really limited" and that "the choices we make can impact on other people's choices…the decision was made that we'd do sustainability in terms three and four". Despite being directed by her principal to engage with SSP, Anita indicated that she greatly valued the opportunity that the program offered for her to personally respond to environmental issues, stating that "I'm interested in those kinds of things and tried to put some of those [sustainability] practices into my life". She noted that due to "being so bombarded by media" about the urgency of addressing environmental issues she would "almost feel guilty" if she did not attempt to teach sustainability concepts to her students.

Despite her desire to contribute to sustainability education, Anita described the SSP professional development sessions as presenting an "overload of information" that was "personally overwhelming" and "quite depressing at times", and which had caused her to question the ability of any one person to make a difference in either the classroom or the community:

> Oh god…what can we do? The issues are so big, and seem to be so far gone, and I guess you also wonder, if I'm here, and I'm making some positive choices and things, you do wonder just how much of an impact that really has in the scheme of things.

Anita related her feelings of helplessness to her observation of daily human behaviour:

> sometimes you go to the supermarket and you see those people who get their plastic bags… and you almost want to go and hit them over the head with something and you sort of think…don't you get it?

Anita described her role in implementing SSP as educating her students about both the finiteness of natural resources and the potential consequences of the decisions that they make about how they use such resources:

> it's trying to get across to the students, right from the word go, about this idea that what we've got on Earth is really limited…[we] need to take into account the fact that things are limited and they're finite…and that the way we use it and what we do with it…can impact on other people's choices.

In light of her own experience of feeling overwhelmed and depressed when learning about the enormity of the potential environmental effects of continued unmitigated resource use, Anita was very insistent that sustainability education through SSP was not "something that you [the students] should…lay awake at night worrying about". She stated that "I think some children would [worry], depending on the way that [an environmental] message was put to them".

Anita insisted that her approach to SSP in the classroom was to "start with where the kids are at and…keep it positive…it's important, but you know, we can't run around like chooks with their heads cut off". Anita believed that implementing SSP with her young students required getting the "balance right" in order to teach at "a level where it's meaningful, but at the same time…not something that's nightmar-ish". She embraced this philosophy by assisting her students to relate sustainability ideas to their own lives and their own actions. She described a learning module in which her students were encouraged to "address the issues of reduce, re-use, recy-cle" and then to relate these concepts to resources, including "electricity, paper and water", and then to "the issue of waste". She stated that she focused on assisting her students to begin thinking about:

> the kinds of things they can do at home, the kinds of choices that their Mum or Dad might make about their lunch box…and whether or not you leave lights on…and even maybe some choices they make about the kinds of toys they might be interested in.

SSP in Practice   Anita shared a double classroom with Robyn to facilitate team teaching. She chose a combined lesson to highlight her approach to SSP. This lesson contributed to a "recycle, re-use, reduce" module which aimed to improve the stu-dents' confidence in: sharing ideas; questioning peers and teachers; and finding dif-ferent ways in which to express their understandings through posters and models. Anita explained that such skills were essential if her students were to be given the freedom to choose how to direct their learning and present their ideas in future years.

Groups of students rotated through four activities based on science and environ-mental understandings established in earlier lessons:

- create a paper windmill;
- play a computer program to identify closed and/or open electricity circuits (this was an extension of a previous class in which the students had assembled circuits using light globes and batteries);
- make a poster by cutting pictures from old magazines to show the ways in which electricity is used; and
- brainstorm ways in which electricity use can be reduced, and make a poster to highlight the message 'turn off the lights'.

The students were obviously comfortable and confident in working in the shared space of the combined classroom, and worked collaboratively and enthusiastically with both peers and teachers. Throughout the session, both Anita and Robyn assisted the students to cut pictures from magazines and to construct their windmills. After 20 min they ushered groups from one activity to the next, irrespective of whether or not the students had completed their task. Several of the students were clearly frus-trated by the need to abandon what they considered to be unfinished work.

Anita noted that she had found some aspects of implementing this SSP module to be quite difficult. She described a lesson that had been designed by Robyn to encourage the students to explore certain aspects of electricity:

> we got some light bulb circuits and let the children have a go at making the circuits to try and get an idea about what it's about, how electricity works, that you've got to have a circuit and that kind of thing…and that electricity can be stored.

Anita described feeling uncomfortable teaching this lesson due to her lack of knowledge of the science of electricity. She felt that teaching a subject such as this placed her in the difficult position where "I have to say something I know pretty much nothing about". Anita's pedagogical approach to this lesson is represented in Fig. 5.1.

Understanding Pedagogy Anita's beliefs about teaching were focused mostly on the need to ensure that students' voices were heard in the classroom. She believed that teachers must consider "children's outcomes", rather than simply direct learning in order to "meet a criteria", stating that "if that's the only reason for doing it, it just doesn't feel like enough of a reason". She suggested that teacher-directed

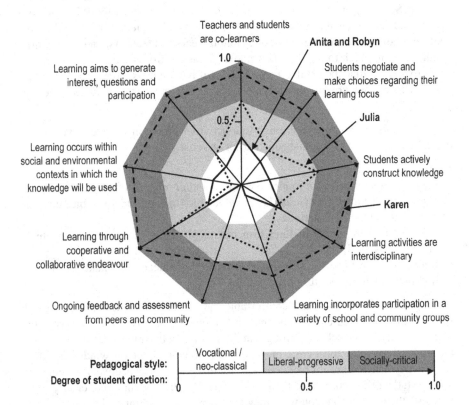

Fig. 5.1 State of play: the pedagogical approach of four teachers, Anita, Robyn, Julia and Karen, to implementing SSP at East Valley Primary School

pedagogies were not only "a bit restrictive", but that they also failed to "take into consideration perhaps a different learning style or a different way of presenting". Anita described teacher-directed pedagogies as being dismissive of children's abilities, because "sometimes you don't know what children can come up with".

Anita indicated that although a student-directed (socially-critical pedagogy) was her "more preferred style", such an approach was only appropriate in classes with:

> students that are a bit older or perhaps a bit more experienced in setting up activities themselves, and are used to the idea of negotiating things [be]cause I think quite a few kids don't actually get the opportunity to do that very much.

Anita was most concerned that such an approach could seriously undermine the confidence and enthusiasm of younger students. She noted that "sometimes kids get ideas and they think it's [going to] be one thing and it doesn't turn out that way…it can be really disappointing I think". As an example, she referred to her anxiety about the depth of the students' interest in the development of tadpoles being kept in a tank in the classroom. She explained that "we do get a lot of enjoyment just out of the fact that they're still here and still alive", and was very concerned about what might happen if the tadpoles had not turned into frogs by the end of the term. Anita explained that it was these types of student expectations that meant that student-directed learning could only work in classes in which the students had developed "quite a bit of trust in their teacher that there was [going to] be a whole lot of follow through", or support, when needed.

Impediments to Socially-Critical Pedagogy   Anita attributed her willingness to trial new pedagogies to the strength of support from her principal. Such support was essential to counteract the negative effects of colleagues, many of whom were experienced teachers, who were openly dismissive of alternative practices and strongly resistant to change. Anita noted that this dissention eroded her confidence and created an unpleasant work environment. In light of this she nominated peer support as the most critical element in ensuring the success of any teacher's implementation of SSP ideals:

> you just really need a really good bunch of people around you, possibly you don't actually need a whole lot more resources…but having people around you that are enthusiastic…I think it's really much more about people you work with…it's not about the resources.

In addition to peer support, Anita acknowledged the importance of a collaborative work environment for developing new ideas, or "people to throw those bright sparks in there", and opportunities to seek advice, or "perhaps get a bit of direction about how to successfully maneuver through some of those things".

In more practical terms, Anita found that the most difficult aspect of changing from a vocational/neo-classical pedagogy was providing space and legitimacy for students' voices. She noted this was often difficult with very young students who are "still very much in that eager to please mode" and still need to learn to be heard, learn how to make choices, and learn how to think more critically about the opportunities presented to them. As a result, she believed that, in practice, student choice needed to be limited in the earliest years of schooling.

5.2.2   *Robyn*

Robyn had been a qualified teacher for just 3 years. She had spent this time as generalist Prep teacher at East Valley Primary School. Robyn was sharing a double classroom with Anita.

Understanding SSP   Robyn explained that her participation in the implementation of SSP was because "I have to do it". She viewed SSP as "a government initiative" related to "trying to save the environment". She explained that the program facilitated "teaching children…how to save the environment" by helping "children to understand how to use less water and less energy". She described the program as a series of learning activities that included "introducing such things as worm farms" and having "children growing their own vegetable gardens".

Robyn hoped that these activities would lead to "a change in behaviour" so that her students would get used to, for example, "using less water…[and] not bringing wrappers to school with their lunch". She described the program as "just teachers and students working together", not only to alter the students' behaviour at school, but in the hope that "at home they can [also] change the habits of parents and older brothers and sisters".

Robyn stated that the program initially seemed overwhelming. She noted that the professional development sessions outlining impending environmental crises had caused her to become "really worried about the world", and hoped that there was not "too much more PD [professional development]…it's too scary".

SSP in Practice   Robyn shared a double classroom with Anita to facilitate team teaching. Robyn, like Anita, believed that her approach to implementing SSP in the classroom was best highlighted by a combined SSP lesson that contributed to a "recycle, re-use, reduce" learning module. She also hoped that this lesson would improve her students' confidence in sharing ideas, questioning peers and teachers, and presenting their ideas in a variety of ways. Refer to the earlier discussion of Anita's SSP practices (Sect. 5.2.1) for a full description of this shared lesson. Robyn's pedagogical approach to this lesson is represented in Fig. 5.1.

Robyn was very proud of the manner in which her students had embraced the underlying principles of the "recycle, re-use, reduce" module, and that through her teaching, her students had found ways in which to become far more environmentally active than most of the teachers would believe. Robyn reported that "in the classroom on a day-to-day basis" her students were beginning to automatically make good decisions about the use of resources and limiting waste. Robyn was most proud of the way in which her students influenced the behaviour of people outside the classroom, stating that:

> we've had a couple of children's parents come in and say 'Oh they've [students] been saying we should do this now and we should do that now'…and there have been a few parents that have approached me and say 'Oh they're telling me to do this, they re-use the back of paper instead of writing [on just one side] and throwing it in the bin'.

Robyn explained that she constantly tried to link SSP ideals into other learning areas, and to facilitate opportunities for her students to put these ideals into action. She recalled, for example, that she had assisted her students to write a letter to a local company who had responded by donating worms so that the students could establish a worm composting farm at the school. Robyn hoped to develop her SSP teaching so that "the community is involved, and children are working with different year levels rather than just being confined to their own classroom". She believed that "hands-on activities" were essential for maintaining the students' engagement, and that guest speakers showed the students "new ways of getting information".

Understanding Pedagogy   Robyn seemed unable to identify or provide an opinion about the critical differences between the pedagogical approaches expressed in the three hypothetical scenarios. She focused on identifying the practical components, or specific learning activities, in each of the scenarios that she had undertaken with her own students, or that her students had experienced as a result of visiting EVNP.

Robyn was unable to decide which scenario best represented her preferred approach to teaching, stating that "nothing came to mind I'm afraid...I can't really say". When asked to choose the scenario that best represented the school's overall approach to implementing SSP, after considerable reflection, Robyn felt confident that:

> children are working with different year levels rather than just being confined to their own classroom and having the community involved and having speakers coming in and hands-on activities...I think we are doing that at the moment, yeah, I'd probably maybe say scenario 2 [socially-critical pedagogy] right now.

Despite being unable to discuss the underlying principles of the pedagogical approaches presented to her, Robyn's comments clearly indicated that she preferred not to employ a vocational/neo-classical approach to teaching. For example, she talked about "having the children doing some hands-on experiments" noting that "we're hopefully going to be doing that soon", and stated that:

> I think you need to go out of the school classroom and then you can come back and bring what the children have learnt there and go from there...it gives the kids such a new perspective and learning.

Impediments to Socially-Critical Pedagogy   Despite overwhelming evidence that her students were highly engaged and developing a range of complex understandings and social skills, Robyn found implementing SSP highly stressful due to the lack of peer support. She valued the strong support of her principal but noted that neither herself, nor Anita, engaged with the other early years teachers who had refused to embrace the educational change process facilitated by SSP. In addition, Robyn was concerned that SSP itself was not sustainable. She believed that "there needs to be more of a whole school approach...I think that if the whole school's involved in things like this there's more of an ownership over the school". She also noted that such a program must be integrated into the curriculum each year rather than timetabled as discrete biennial subjects as had occurred at East Valley

Primary School. Robyn feared that the momentum for educational and behavioural change that she felt she had helped to build would quickly diminish if it was not continuously supported.

5.2.3 Julia

Julia had been a qualified teacher for just 5 years. After 4 years as a generalist Grade 5–6 teacher at East Valley Primary School, she had been asked to teach Grade 1–2 for the first time. Julia was also responsible for managing the Junior School Council.

Understanding SSP Julia was unsure of the overall purpose of the SSP: "I know very little about sus[tainability] and I get…all confused…I know we're trying to be sustainable here…but being a sustainable school and what that involves…I'm not that clear on what that is". She did understand, however, that SSP was being used as a vehicle for change, stating that her principal "wanted to point us in a direction" because "we were floundering". She explained that central to this change was the notion that the specified learning outcomes of a curriculum document could be achieved through concepts and topics identified by students. She described this as an "inquiry" approach that had been identified by her principal as an appropriate alternative to the teacher-directed pedagogies that had become well-established throughout the school. She described those established practices as:

> like a blanket…all the grades did the same thing each week…we didn't even try and bother to get the information about where the kids [were] at, we just said okay, we're going to teach you all about the life cycle of a chicken whether you know it or not…every week we'd have the worksheets we'd give them.

Julia explained her principal's approach to facilitating change: "she's made us feel like we've had a say in it too…we brainstormed…to map all the things kids need to learn…to be responsible [citizens]". All of the brainstormed ideas were then sorted into "four curriculum organisers…one being environmental sustainability". She noted that her principal had obviously "had a plan that ended up with sustainability…and she pushed us, well…led us in that direction". Julia explained that the teachers developed statements through which they could focus student learning for each of the curriculum areas at each grade level. She believed that the Grade 1–2 statement for environmental sustainability, "The way resources are managed through reduce, re-use and recycle." was excellent because it enabled her to incorporate concepts that her young students could easily "relate to their personal lives", and relate to "their local area…not just some other country or out whoop whoop [region] that they're not related to".

Julia described her role in the implementation of SSP as teaching "an integrated unit" that focused on "reduce, re-use and recycle". She understood the aim of this unit to be "making kids aware" of the environment and "just not being wasteful". She believed that this required the students to:

learn what those three focuses are, and the concept of what waste is and that we want to reduce waste because we are creating too much waste…we focus on electricity, paper and waste, and it's a good thing to reduce, re-use and recycle to reduce our waste.

SSP in Practice Julia readily accepted that she was neither the sole expert, nor always the sole authority in the classroom. She chose a lesson about energy, held at a neighbouring secondary school, as an example of her commitment to implementing a socially-critical pedagogy. This lesson involved collaborative student-directed learning through multi-age student–student interactions. Julia's Grade 1–2 students worked in small groups that rotated through five practical and exploratory activities that were planned, supervised and explained by Grade 9 students:

- Static electricity—students rubbed large balloons with various materials, such as sheepskin and velvet, in order to create static electricity that enabled the balloons to stick to the wall and make their hair rise. They wiped a glass rod with material to create a static charge so that the rod could cause a stream of water to bend;
- Draw something about electricity;
- Create an electric circuit using globes and batteries;
- Charge a solar panel in the sun so that it could make a light bulb glow; and
- Investigate the workings of a shake torch, in which a magnet moves up and down inside a coil to make the torch glow.

The entire lesson was conducted with little intervention by either Julia or the Grade 9 teacher. Throughout the lesson all of the students were engaged in activity and associated conversation—there was a high degree of knowledge sharing and language development through student–student communication. Julia commented that this collaborative lesson provided access to equipment and the knowledge of a science teacher, filtered through the Grade 9 students, of a topic that she was not confident to teach. The Grade 9 students reported that, prior to the lesson, they had worked hard to confirm their own understanding of the activities so that they could confidently explain the science of electricity, and answer any questions posed by the younger students. Although the Grade 9 students were the 'experts' during the practical part of the lesson, the younger students themselves became 'experts' during subsequent classes in which they provided descriptions and explanations to both Julia and their peers. Julia's pedagogical approach to this lesson is represented in Fig. 5.1.

Julia also described lessons in which she had attempted to modify her practices in ways advocated by her principal. For example, she described undertaking a "tuning-in" lesson before undertaking a unit about the water cycle:

I found out what their [the students] misconceptions were and I found out where their interests lie, and what they know about. For example, my guys [the students] knew quite a lot about the water cycle, so while that was one of our essential things, I didn't spend a lot of time on it…whereas my guys didn't really know a lot about the journey that, for example, paper made…and the journey that toys [made]…they had no idea [of] a lot of products, where they came from, the journey that they got to be made, and then what happened to them when you finished with those products.

Julia acknowledged that her usual approach would have been to complete the water cycle unit as planned, but after undertaking this tuning-in lesson, and with the support of her principal, she introduced an activity in which some of her students:

> went into the opp[ortunity] shop down the road here...so we all donated products and sorted them into departments, and the kids sent out invitations and we had it open...so for the whole day the kids had their departments to man, and we had all the Prep, [Grade] 1–2 children come over with fifty cents, and we had parents and the opp[ortunity] shop people come.

Julia explained that this not only assisted her students to understand the possible life-cycle of a toy, and aspects of recycle and re-use, but was multidisciplinary in that the students were also required to work with members of the local community, as well as manage customers and the exchange of money. She noted that, as a result of implementing SSP, "we're at the beginning of getting more of the community involved, it is a fantastic thing...we've got kids, Preps to [Grade] 6's, doing all sorts of things...across the school".

Understanding Pedagogy Julia indicated that, through implementing SSP, she had developed the understanding that content and pedagogy contributed equally to educational outcomes, and that pedagogical change did not necessarily require more time or significant curriculum change. Although she found the change from a traditional teacher-directed approach to what she described as an inquiry-based pedagogy to be personally demanding, she had observed that when her students chose "their own focus for study" and engaged in hands-on learning "rather than just sitting there passively" they did "learn more and respond better". She explained that she kept these observations in mind when assessing different pedagogical practices.

Julia noted that, despite the apparent success of the pedagogical changes she had introduced as a result of implementing SSP, such student-directed learning had to be carefully managed and referenced against the relevant government curriculum, as:

> you still want to make sure they've [the students] covered certain stuff, [be]cause their area of study might be out of left field, completely unrelated to what they need to know...there's some stuff that you have to teach them...the non-negotiable things that you have to teach the kids...the essential knowledge...and essential skills and essential behaviours.

Julia believed that, irrespective of the learning focus, students should be encouraged to negotiate different methods for presenting their understandings because this gave students "an opportunity to show their skills through their strengths". She also thought it would be "good" for students to receive "peer feedback" or assessment "not just from their teachers, but from their local community". She admitted that she had not yet found ways in which to fully accommodate this idea in her own teaching.

Impediments to Socially-Critical Pedagogy Julia explained that implementing SSP had been "a steep new learning curve". She acknowledged that her principal's "knowledge has been really valuable" and that the teachers at East Valley Primary School had "been supported by a lot of very good professional development" without which they "would have been lacking a lot more knowledge". She explained that, in addition to the professional development "we've bought heaps of books" to assist the teachers with implementing SSP, but that other resources were limited, such that, for example, "we've done electricity this week and there's only one electricity tub, so we're sharing it between about twelve grades". Despite the obvious lack of classroom resources however, Julia indicated that her approach to implementing SSP would benefit most from improved access to safe and secure outdoor learning spaces, and the development of opportunities for her students to work collaboratively with parents and other members of the local community. Julia noted however, that these aspects of a socially-critical pedagogy were potentially difficult due to specific characteristics of the community in which her school was located. For example, she explained that although she would like to develop learning projects within the school grounds, she was concerned that her efforts would be compromised by vandalism:

> I'd love to have a veggie patch out here, and the chickens…but a concern is funding it [and] then keeping it safe…I'd be so nervous if we did that here we'd have vandalism and stuff. You'd set up all those wonderful outdoor things and you'd have to somehow hope that the community would respect it…my fear is that if we set up a wonderful area it'll get trashed.

Julia acknowledged the advice from CERES that "when you build these things the community gets behind it on the whole and it doesn't get wrecked because they support it and value it because they have ownership of it" but continued to be concerned that vandalism would "be so disappointing" and would waste "the initial outlay of money".

Julia also indicated that it was difficult to create opportunities for her students to learn through activities undertaken by "multi-age student groups" or in collaboration with members of the local community, or even the students' parents. She hoped that her school could establish a program that enabled the students, particularly those in grades 5 and 6, to "do a lot of work with the [neighbouring] high school", but noted that, as there had been quite "a bit of resistance" from other teachers within the school "that may or may not go ahead depending on numbers…there's different people with different opinions". Similarly, when considering community involvement in potential SSP projects undertaken with the students at the school, Julia rather sarcastically commented "weekend workshop…that'd be interesting to get parents to show up!"

5.2.4 *Karen*

Karen had been teaching for 26 years, the previous 13 at East Valley Primary School. As an Information and Communication Technology (ICT) educational specialist, Karen was responsible for network administration and ICT education across all grades in the school. Three years prior to the implementation of SSP a shortage of classrooms at East Valley Primary School forced Karen to look elsewhere for an opportunity to develop an ICT teaching facility. She assisted the school to negotiate an agreement for the use of a vacant training room within the grounds of the neighbouring East Valley Nature Park (EVNP). Initially, small groups of students attended for just a few classes each term for ICT lessons. Since then, and in collaboration with EVNP staff, Karen had developed an environmental multi-literacy program which contributed to the school's implementation of SSP. Each year, every class attended EVNP for at least two 1-week periods. Normal classes were suspended during these weeks because all learning activities were undertaken at EVNP.

Prior to implementing SSP at EVNP, Karen had taught only ICT. She indicated that implementing SSP had required much learning because she had never before been involved in teaching environmental education.

Understanding SSP   Like her principal, Karen viewed SSP foremost as a teacher education program. She believed that SSP focused on "teaching teachers to empower children to understand their environment", so that teachers can assist their students "to learn about their environment, to have respect for their environment, and to actually act on that, and therefore, in the long term make changes to the world—starting with their world". This appealed to Karen's strong desire to make "a difference", and underpinned her motivation for teaching, which she described as "trying to teach the word respect…respect for themselves, respect for the environment…if I can impart that to children…then I have made a difference".

Karen believed that in order for students to feel respect for another species, they must first "understand that species", and that this understanding can only be achieved through an authentic learning context. She believed that it is only after the development of this understanding, and therefore respect, that learning can progress beyond the most simplistic level of reciting facts to become self-questioning of attitudes, beliefs and behaviours. In facilitating this, Karen acknowledged the influence of her own learning from nature—"I've actually seen how they [animals] work as a team, and that we can learn from them"—as the motivation for developing a truly cooperative and collaborative relationship with the community of people at EVNP.

SSP in Practice   Karen chose the first day of a 1 week visit by Grade 3 students to her classroom at EVNP to highlight her approach to socially-critical pedagogy and SSP. Superficially, Karen's classroom seemed like any ICT learning space incorporating a bank of computers and student chairs. However, there were no bells or set times for rest breaks at EVNP. The students were expected to work cooperatively and manage their time, not only to complete their tasks, but also to ensure they were present for specific activities, such as guest speakers. Each student chose when to eat or rest, according to the demands of the activities in which they were involved. Furthermore, due to the lack of facilities at EVNP, every student was required to prepare their own lunch and to clean-up when finished. The Grade 3 students accepted these responsibilities, and there was no evidence of student misbehaviour nor the necessity for a more rigid routine.

Early in the morning, the students undertook a short computer lesson designed to develop their proficiency with a new multimedia program so that they could create a visual diary of their work with one animal at the park. Karen showed the class an example of a completed 'working with an animal diary', and then provided a step-by-step guide to demonstrate the basic operation of the multimedia program. Although some of the students carefully followed Karen's instructions, others simply explored the program's capabilities, listened in for the occasional tip, or asked direct questions as required.

During the afternoon, the students were introduced to an "expert" visitor, the local "bee man". Karen organised this visit due to recent media reports, noticed by her students, concerning the plight of bees due to the lack of nectar during the continuing drought. She noted that visits by community people provided valuable educational opportunities, because "the children actually get the message not just from a teacher but from other people...it is really powerful...[when a message comes] from more people than just the teacher". The visitor brought with him important components of a beehive: a set of beekeeper's clothing and tools; several types of bees in hand-held perspex cubes; and various materials including bee's wax and pollen. The students were encouraged to look, touch and ask questions. It was this latter aspect of the lesson which highlighted the benefits of Karen's pedagogy. The quality of the students' questions indicated significant critical thinking, and reflected learning from previous experiences at EVNP. For example, the students queried the possibility that the length of the usual life cycle of bees might change in response to environmental change, as had occurred with some of the animals they cared for and studied at EVNP. Karen's pedagogical approach to this lesson is represented in Fig. 5.1.

While attending the park, the students were responsible for the daily care of various indigenous animals. In addition, the students assisted rangers and researchers with special activities, such as conducting biological surveys, or assisting with the artificial insemination breeding program for endangered birds. These experiences successfully and deeply engaged the students. Many of the students brought their parents to EVNP after school and on weekends, and they had been observed explaining aspects of the endangered species breeding program to strangers, and asking visitors to pick-up their rubbish. The students had also raised funds to assist various projects at EVNP. For example, Karen described the students' interest in a pond redevelopment project for the Musk Duck breeding program: the "junior school council got together...they wanted [their fund raising proceeds] to go to the Musk Ducks so they could breed", because they thought that it was "really, really important... that these ducks get together to breed". It was evident that the students' contributions had become an integral part of the daily routine and special projects at EVNP. In response to the high level of student interest and involvement, a weekly postal service between the school and EVNP enabled the students to correspond with rangers, not only to stay informed about aspects of EVNP in which they had developed a personal interest, but also to seek answers to questions for projects undertaken in other classes.

Although all of the teachers at East Valley Primary School were encouraged to attend EVNP with their students, they rarely attended more than once. A Grade 3 teacher who, for the first time, accompanied the students to observe their first day at the park, was openly uncomfortable with the relative lack of structure, both in terms of classroom management, and the degree of freedom given to the students to move around EVNP as they completed their tasks. Karen believed that most of the classroom teachers found their experience at EVNP to be "extremely threatening", and often formally complained to their principal about the lack of rules, flexible lesson times and inconsistent teacher supervision. She did not think that the majority of teachers understood that by encouraging "children to work to their own time...this doesn't seem like school work...it's actually fun and they want to stay".

Similarly, student learning at EVNP was not easily described in terms of traditional "tick-the-box" curriculum outcomes. Karen found this type of education difficult to justify, particularly in relation to assessment. She stated that others did not understand that "this program is not meant to be an end in itself, it's meant to be a springboard of changing attitudes for children to take back to East Valley Primary School". She admitted that "I don't know what I can truly measure here...maybe... the measurement should be back there [at school and at home], of children and their attitudes...and their behaviours".

Understanding Pedagogy Karen had a very specific idea of what constitutes effective and worthwhile education. She indicated that the most essential factor in the development of an effective teaching–learning environment is the degree to which students are able to direct their learning. She commented that a traditional "teacher-controlled" vocational/neo-classical pedagogy is "just simply not as powerful as coming from the children" and "I don't like this total control". She explained that:

> I've sat in too many meetings over the years where people are designing curriculum...so that it fits into 2.3, ahh okay yes we've done that now let's cross it off...let's move on to the next section, right, now 2.7...I guess it's, to me, just superfluous, it's just ridiculous to try and teach that way...it has no meaning, it has no meaning to the teacher because they're doing it to please somebody who's written this curriculum...therefore it's not going to have any meaning...to the child.

In comparison however, Karen supported socially-critical pedagogies for their ability to provide a more holistic approach to learning, not only for students but also for teachers. Her ability to determine the basic learning direction provided a degree of motivation and job satisfaction, and addressed all major educational goals without a prescriptive 'tick-the-box' approach. She strongly believed that any effective educational journey begins with ideas, interests and needs which "come from the students...ground swell, there's nothing so powerful as ground swell". The importance of student "ground swell" was central to Karen's curriculum planning. For example, she had recently obtained materials for a worm farm as an SSP initiative, but had not yet begun construction because there had been little student interest. Karen stated that "I'm showing the kids where it is, and we have a little bit of a talk about it" but it will not go ahead until "a group who are really keen and really interested" want to take on the project. This exemplified Karen's "preferred...child-centred and child-directed" socially-critical approach in which learning was a "negotiated" process. She noted that she provided "a fair degree of freedom" and "choice" because "I don't expect children to all have a high level of ownership for everything they do, that's just ridiculous". Similarly, Karen encouraged all of the students to "direct their learning" journey within all of the experiences and opportunities available at EVNP. She explained that:

> some children will go off on a total different tangent right, and I find every week, of every group that come here, we go in a different [direction]…even though we have…components that we will look at for the week, we all end up going in a different way.

This type of authentic learning at EVNP was exemplified by the students' experiences of undertaking biological surveys with an aquatic researcher. Karen described instances in which the students:

> all end up in waders out in the water…some will come back and want to really go on with the microscope…others won't be interested, they'll want to come back and they'll take cameras and they'll want to do a movie about it…others will just want to work with photographs…others might…create an animation in macromedia…how they come back from that is really a personal thing…It's up to the children…where they go with the information that they've been given…if they have done a certain topic, if they want to then go back and find the ranger, they just go round [to] the office, knock on the door, excuse me, got a minute, and this happens all the time.

Despite the lack of peer support for her work at EVNP, Karen noted that she had significant support for the program from parents. Even though the students working at EVNP were not shielded from the less pleasant aspects of the real world environment, including, for example, the death of animals that they had perhaps raised from birth, and that the students often returned home with muddy or torn clothes and with scratches and insect bites from their outdoor activities, the parents respected the high level of engagement and learning offered by the EVNP experience. There was considerable community support for Karen's work, particularly as a result of a significant decrease in vandalism in the park since the beginning of the program.

Impediments to Socially-Critical Pedagogy Karen believed that the most difficult aspect of her role in SSP was the organisation and time it took to establish community relationships and involvement in the educational process. She found that:

> in actual fact you've got to drag people in…in a lovely world these people would come knocking on your door and say 'I have these wonderful skills and if you need me, I'm available'…but the world doesn't work that way, I find we have to go and seek them out.

Karen stated, for example, that "it's taken me two years to get somebody to do some indigenous stuff, two years…contacting cooperatives, going round in circles trying to get help". However, she saw these challenges as more than just her role as a teacher, but a way of approaching life, because "you really have got to work on it…if you're going to sit around in life expecting it's going to come on your doorstep it probably just won't happen".

5.3   Mountain Primary School

Mountain Primary School was a co-educational government school which catered for almost 550 students. The school was located approximately 500 km from Melbourne in a residential zone of a large multicultural provincial centre with a population of almost 45,000. This centre provided a variety of food processing facilities in support of nearby agricultural activities. The effects of a severe drought however had resulted in significant unemployment in both the agricultural and processing sectors of the community.

Mountain Primary School incorporated many modern buildings, designed to make use of spacious school grounds by ensuring that classrooms had direct access to shaded outdoor learning areas. Despite the effects of several years of drought, the school had found ways in which to maintain a variety of native and cottage-style gardens, as well as an award winning kitchen garden. A variety of public amenities, including a small community shopping centre, municipal playgrounds and sporting facilities, and several types of natural landscapes including a river, natural bush land, and a large flora and fauna reserve were located within walking distance of the school.

An important component of Mountain Primary School's vision was to prepare students for their future. The school embraced the idea that the best way to assist the students to acquire the knowledge and skills they needed for their future was to create a learning environment that incorporated a wide range of supportive partnerships between the teachers, students and their families, and the local community.

One teacher, Elizabeth, and the manager of the school kitchen garden, Stephanie, shared their experience of working to implement SSP ideals at Mountain Primary School.

5.3.1   Elizabeth

Elizabeth had been a generalist primary school teacher for 30 years, the previous 8 of which had been at Mountain Primary School. She was teaching Grade 3 students as well as coordinating yard duty responsibilities and the delivery of the language other than English (LOTE) programs. Elizabeth stated that she had previously taught environmental education only as "part of general curriculum as needed", but often designed LOTE programs to promote environmental activities in the school.

SSP Implementation   Despite the fact that Mountain Primary School had been a SSP five-star accredited school for more than 2 years, neither the principal (undertaking her second year at Mountain Primary School), nor the majority of the classroom teachers were aware of the program or its ongoing status within the school.

The implementation of SSP had been instigated by the previous principal, who, according to Elizabeth, had a personal interest in environmental issues. Elizabeth explained that although the previous principal had made her responsible for the implementation of SSP throughout the school, her involvement with the program was neither voluntary nor enjoyable, and she was adamant that she "wouldn't have chosen to" participate because it was such "a huge task". She noted that although she had participated in professional development sessions held by CERES at the school, she had not been able to attend external training sessions: "I applied to go to one at the zoo, and I was going to take a workshop, a full workshop, but then various things arrived at the school...it would've just been too hard to go there".

Understanding SSP   Elizabeth viewed SSP to be "not just a school curriculum". She believed that while it "starts in the classroom with education...it's just not a 9 to 4 concept...it's a 24-hour concept". She stated that "I see it [SSP ideals] as moving up throughout the school, with [the students] having bigger jobs, being more aware, [for example] start off in the classroom with them [the students] being more aware of their playground and then the bigger playground". She described SSP as "a community...outreaching program", because the students were encouraged to take their new understandings "back to the family", and Mountain Primary School had become a working model for how a school could engage with environmental sustainability. Elizabeth explained that when implementing SSP:

> I want them [the students] to have a core understanding of the importance of whatever that aspect [learning topic] is, but I also want them to be able to repeat and teach someone else about it, because it's not until you can tell or teach or impart that knowledge that you actually know it yourself, and that you know you know it.

SSP in Practice   Elizabeth explained that her role in environmental education, as it related to the implementation of SSP at Mountain Primary School, was concerned mainly with the organisation of waste management and recycling, and water and energy saving activities rather than classroom teaching or student-directed action or activities, stating that "logistically it's full on". She described her rubbish collection routine as an example of the how the implementation of SSP had influenced Mountain Primary School:

> on a daily basis there are the bins to put out, collect in...weekly bins to put out, collect in, fortnightly nineteen recycle bins to collect and put out...re-name them, because one thing I did was put big names on all of the bins and they constantly get taken off, so when the recycle bin has lost its name then I have six kids in my grade saying 'Oh where does that one go?'.

Elizabeth resented the effort and time this routine took, explaining that "to physically go and get all the recycle bins, line them up, and walk them out to the front of the school, because the kids can't do it unsupervised, takes up twenty minutes minimum". Only Elizabeth's class undertook the responsibility for this routine, because Elizabeth felt that "it would be horrific to have various classes do it" because "it has to be well organised otherwise it would really fall in a heap very quickly". She believed that:

> if you had one person looking after junior bins and one person looking after middle school bins and likewise for the senior school you'd still have to have someone overseeing it, and

if you happen to have two of those supervising teachers away on one day, then you'd have to...go through the explanations with CRTs [casual relief teachers], and that would be horrific.

Elizabeth was proud that the time and effort she invested in establishing and maintaining the waste management system had greatly reduced the volume of rubbish generated in the school. She attributed this success to the reward system she had introduced with both "individual acknowledgement and class acknowledgement". She explained that each week "every child who has a waste free lunch" went into a draw for the "lucky lunchtime lotto" prize, while the "golden wheelie bin" was awarded to the class with "the lowest amount of waste".

Despite the apparent success of the waste management system, Elizabeth was frustrated that the classroom award was usually won by her own class, and that continued improvement from the rest of the school was not forthcoming. For example, although the Grade 3 students had observed that many bins were only half full and suggested that fewer bins be used, Elizabeth had resisted, stating that a general lack of student initiative meant that even the smallest changes required "a bit more education...well before it happened [so that] all the classes knew why it was happening and what was happening".

Elizabeth was unsure of the status of SSP in the school beyond waste management, as the accreditation process had been accomplished through specific special projects which ran for a limited time. For example, she remembered that during the:

> year before last...we had a fairly big energy-wise education program, and we had an energy monitor for each room, and that person was in charge of turning off all the switches each night. This year we haven't to my knowledge. I don't think we've had any special focuses on energy.

Elizabeth believed that education regarding the sustainable use of water was probably being undertaken because "we all bring water education into our environmental awareness, because of our location [a drought affected region], and so I would imagine that all kids here are fairly water-wise as a matter of course". She also referred to the role of the kitchen garden for providing water education, because the kitchen garden manager (Stephanie) had "done a lot of...research into...some of the sprayers coming into the kitchen garden" and that Stephanie "imparts that knowledge to the kids" and "because so many kids use the kitchen garden it's sort of a natural filtering of information". Elizabeth's pedagogical approach to implementing SSP is represented in Fig. 5.2.

Elizabeth also explained that the learning and teaching of SSP ideals at Mountain Primary School also occurred in lessons during which her students visited the kitchen garden. Each school term, Elizabeth's students spent several hours working in the kitchen garden with Stephanie. Stephanie, a landscape design expert, had designed and constructed the garden, and continued to maintain the garden for the school. The garden was protected by a high wire fence and locked gate so that students could only enter the garden when she was present. Stephanie allocated tasks to the students as they arrived, in groups of six, for half-hour long garden maintenance sessions, and closely monitored their progress. Tasks included, for example, the weeding and transplanting of seedlings that Stephanie had propagated. Stephanie

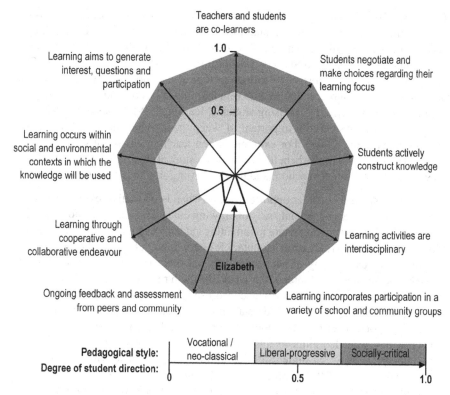

Fig. 5.2   State of play: the pedagogical approach of one teacher, Elizabeth, to implementing SSP at Mountain Primary School

explained that she tried to maintain a very seasonal approach to the activities she gave the students in order to assist them to develop a feel for the changes and cycles of growth. Most of the students worked in a matter-of-fact manner, appreciative that garden maintenance seemed less like the lesson that they had come from, but with seemingly little real enthusiasm for the task at hand.

Stephanie noted that she had developed a good relationship with Elizabeth's Grade 3 students as they worked in the garden each week, and that this had helped to reduce vandalism because the students did not wish to hurt her feelings. She also noted that she was most careful to shield the students from the more unpleasant realities of life that the kitchen garden might reveal, such as the death of one of the chooks or an injured bird.

While the Grade 3 students were working in the garden, Grade 5 students arrived in order to collect samples of different types of leaves for making pencil rubbings to help them describe different leaf structures, and a Prep class teacher used the garden for a measurement activity during which the students used icy pole sticks to measure objects of their choice. This teacher stated that she had no knowledge of SSP but that the garden was a convenient and pleasant space for outdoor activities. At no time was there any interaction between the Grade 3 students and the students from other classes.

Understanding Pedagogy Elizabeth's approach to teaching was founded on a strong belief of the appropriate manner in which teachers and students should interact, and of what constitutes an educational context. Elizabeth strongly preferred a school-based, well-structured and teacher-directed approach to learning. She believed that it was important to follow a "logical and progressive approach" in order to best take "into account the different learning styles of students" and to provide a variety of ways of "assessing students' learning". When designing learning units, Elizabeth considered it important to "keep the activities in school where the actual task is going to be". For example, when teaching about native plants, she believed that it would not be beneficial to take her students to a native forest "because they haven't got a forest at school and they're never going to [have one]".

However, she did consider opportunities for "hands-on [activities] and working with a partner" to be a beneficial "way of drawing out and pushing out and maximising" students' understanding while maintaining student "engagement". She believed this to be particularly relevant at Mountain Primary School, because the "low socio-economic" status of many of her students meant that they needed a firm push to move beyond a superficial interest, or what Elizabeth referred to as an "I like doing that" attitude. This encouraged Elizabeth to use knowledge-sharing activities as a way in which to develop "core understanding" and confidence in her students. She provided opportunities for her students "to repeat and teach someone else about" concepts they were learning, in the belief that the "imparting of knowledge and communication" between students during knowledge-sharing activities enabled her to judge "the extent of the learning and the extent of the success" of her teaching. Elizabeth found that such opportunities depended not only on the willingness of teachers to work together, but also practical issues such as the location of classrooms, stating that "if you've got a room that's fairly close by" and "it's quick and easy to get to that room, you can do it".

When asked to consider socially-critical pedagogies, Elizabeth agreed that they seem to address the "interpersonal and interrelated curriculum" requirements of Victorian state government curriculum guidelines, particularly in terms of opportunities for students to design their "own projects" and assess "their own personal development". However, she believed that although such pedagogies "would be very easily managed in the secondary [school] surroundings" where they "would fit in very well with the sorts of subjects they do" they are most inappropriate for primary school students. Elizabeth believed that socially-critical pedagogies failed to engage "most of the students for most of the time", and that it was not possible to "have primary schools going out into the community on a regular" basis. She explained that "interaction with the community" at Mountain Primary School occurred only during special events, such as an impending mini-fete, during which the community could enter the school grounds in order to purchase items from the kitchen garden.

Impediments to Socially-Critical Pedagogy Elizabeth considered the lack of "resources" and limited "funding" to be the most critical limiting factors to her teaching. She believed that the lack of specific teaching resources, such as science equipment, restricted the types of learning activities that could be offered by the school. She indicated that access to resources and expertise outside the school was limited by the need to minimise student costs when organising excursions. She believed that with extra funding "we could take them [the students] to CERES, the Collingwood Children's Farm, Botanical Gardens in Melbourne, there's any number of places you could go, but, I'm guessing, [it would cost] probably 600 to 700 dollars for a bus". She also commented that excursions needed to be organised at least "twelve months ahead" of time, stating that "we're doing budgets for the whole of next year now...so we really have to decide now". She noted that due to the low student fees at the school, each student could attend only two excursions each year, and that there was always significant debate amongst the teachers as to where each of these should be. Despite this, Elizabeth was confident that these short falls did not significantly influence her classroom pedagogy, and in no way inhibited the "sharing of information and sharing of learning, and designing student-centred classroom tasks". Elizabeth compared her personal perspective on school resources to the notion of "sustainability in environmental education" which supported the idea that "you use well the resources you've got, not go out and pluck new resources".

5.4   Ocean Primary School

Ocean Primary School was a co-educational government school attended by 500 students. It was located in a predominantly professional, middle-class residential suburb approximately 30 km from Melbourne, with a local population of about 6,000. Ocean Primary School was built on spacious grounds that incorporated a range of facilities, including dedicated sporting grounds, grass paddocks and native bush land. The students were encouraged to use all of these facilities, and to participate in the running of a small school farm. A variety of public amenities, including a large shopping centre and municipal parks and gardens, playgrounds and sporting facilities were close by. Several types of natural landscapes, including a wetlands reserve, were located within walking distance of the school.

The school prided itself on facilitating the development of a broad range of teaching and learning strategies in order to accommodate the needs of all students. Ocean Primary School had been a SSP five-star accredited school for 12 months.

One teacher, David, shared his experience of working to establish SSP at Ocean Primary School, and to incorporate SSP ideals into his classroom teaching.

5.4.1   David

David had been a senior Grade 3–4 teacher at Ocean Primary School for "too many years". In addition to managing the teaching and learning program for Grade 3–4, David was a staff mentor and SSP coordinator. He had extensive experience in teaching environmental education through the core curricula of both science, and studies of society and the environment (SOSE).

Implementing SSP David explained that the principal of Ocean Primary School had made the decision to implement SSP in the hope that the program could address several "issues identified as school problems". Some of these issues reflected concerns about the ageing state of the school grounds. David explained that the idea to develop an environmental program:

> started with the parents wanting to do something with this [courtyard] area. It was an environmental disaster. It had gravel. In summer it'd smoke up into dust, in winter it would flood, and so they wanted a nice area, we wanted a quiet area.

David noted that this coincided with the need to "re-do" other aspects of the school grounds, because "I think the school council got cold feet about limbs falling down [so] they cut down all the big pine trees". He explained that, compared with other schools, Ocean Primary School was situated on:

> a very large site...so we felt we had to be environmentally responsible about what we did with it, that it was a learning area...a learning area isn't just a classroom, we had all these wet areas, but we wanted other special learning areas, so now there's the indigenous garden, there's the [courtyard], there's the farm, there's a few others we've got planned.

In addition to addressing the need to improve the school grounds, David explained that SSP had been identified as a program that could respond to observations by both the parents and the teachers that many of the students lacked basic life skills and "expected to be waited on hand and foot". In order to assist the students to develop some independence and decision making skills, the school introduced a campers program for all of the students in grades 3–6, held in nearby bush land. David explained that because these camps were held in an "environmentally nice place...we talked about the environment", and the students participated in a range of conservation activities linked to the sustainability ideals of SSP.

David also explained that a significant issue for Ocean Primary School was the high number of students, mostly boys, who found it difficult to engage with learning, and whose preferred learning styles were rarely supported. He described:

> kids that just don't want to learn...you know that they're really nice kids...some of those kids are the ones that cling with you on yard duty and want to talk to you...they've got a totally different agenda, but they can't help it, that's their learning style.

Understanding SSP David believed that SSP provided an opportunity for the school to become "environmentally responsible" for their bush setting, while addressing the teachers' observations that the students lacked self-sufficiency, and had difficulty determining appropriate behaviour in unfamiliar situations or difficult environments. He stated that SSP was a vehicle through which to develop the students' ability to contribute to society in an active and responsible manner, because environmental learning is "not a subject, it's something we teach because we teach life" and should be part of "our social responsibility and our civics and citizenship work". He explained that his personal involvement with SSP came from his understanding that the program offered more than simply addressing issues related to the natural or outdoor environment. He stated that:

> for me personally, it was also environment in a more global way, [be]cause I also talk about personal environment to do with the way kids relate to each other, how they work in group work, how we don't have to be best friends with everyone but we need to know how to get on with people and cooperate.

He believed that SSP helped to achieve these goals by broadening the notion of where and how learning occurs, identifying the school grounds and community spaces as "authentic learning platform[s]" from which shared experiences became authentic contexts for exploring life as well as important curriculum outcomes in literacy and numeracy. David explained that the addition of a variety of different outdoor learning areas and associated activities had not only "helped motivation in the classroom" but had also improved student behaviour so that "there's not many control issues". He explained that the SSP initiatives had assisted the teachers to engage the students who now:

> want to give up their play time to help in the farm...they want to be there, they're happy doing the compost and they can talk about it and you see them as a different kid. I see their old teachers in the yard and they say 'What have you done with so-and-so?'

David identified the most essential learning outcomes of SSP as "behaviour changes for themselves [the students] and for the world", and noted that change was most important due to the modern lifestyle: "it's not a sustainable way of life...our kids and grandkids can't have it like that". He explained:

> I find it [SSP] addressing a pendulum that swang too far in the 60s, 70s and 80s, you know there was a time when there were no supermarkets and there were no plastic bags and I think people have forgotten...if you look at modern history, even just since European settlement in Australia...the way people lived in this environment for about two hundred years... it's [modern life] been shown to be not sustainable.

David viewed the need for behavioural change as being "as much to do with our selfish environment...having it nice for us". He used the example of supermarket plastic bags: "these plastic bags are all over the environment, that's garbage, I mean that's not the way to treat ourselves and each other". He explained that this attitude should not be considered radical or "suspicious" because "we're not out there hugging trees all day...we dress conservatively you know", and that having a "nice environment" was a worthwhile goal, "even if it doesn't affect global warming".

In addition to improving his students' ability to contribute to society, David believed that he had a "personal incentive" to implement SSP in that it satisfied his "altruistic" reasons for teaching. He described this as:

> building better pillars of society...leaving the community in a better way than you found it, not just taking up the environment and wasting space or using up the air, but actually contributing, actually making a difference.

Similarly, he considered his role in embracing SSP as a way for him to contribute to the teaching profession, stating that:

I have to do it in the classroom and I've got to try it so that if it works for me it can for others...if I can try it out then I can say to other teachers 'oh, this works really well' or 'you might want to try this this way'...so there is that professional aspect.

David acknowledged that not all of the staff at Ocean Primary School were interested in engaging with SSP. Despite this, David aimed to share the development of SSP with all teachers, stating that it was important to:

explain to the rest of the teaching staff, maybe at a staff meeting, what you're doing and why or what the motivations are, even if their classes aren't involved, because the facility that is there as a result of having done it will be shared by all the kids eventually, somehow, and also it helps if all the staff have ownership of it and know what's going on.

SSP in Practice David believed that neither the concept of sustainability, nor the learning outcomes of SSP, were discrete topics or units to be taught in isolation from other aspects of life, but rather sets of ideals that were "meant to be cross-curriculum" in focus.

David chose a lesson during which his students were undertaking an investigation in order to write a newsletter to inform parents about the environmental status of the new courtyard to highlight his approach to socially-critical pedagogy and his teaching of SSP ideals. David achieved an authentic integrated approach to SSP by incorporating the students in the ongoing management of their school environment. David's classroom was one of several that faced a newly constructed courtyard of walkways, planted trees and shaded seats. The students were assessing the use of the courtyard in terms of: when, where and what type of litter was dropped in the yard; who dropped the litter; why they had dropped this litter; and what could be done to change the litter-dropping behaviour. David explained that this project integrated the SSP ideals of waste management, particularly in terms of reduce, re-use and recycle, with skills developed in: science (process of investigation); mathematics (the representation of data); literacy (writing informative letters to parents, teachers and students about their findings); and social studies lessons (developing and implementing an action plan to change student littering behaviour).

David's classroom resembled a flea market in that the afternoon's project materials were piled haphazardly upon tables positioned apparently randomly around the room. The classroom doors were wide open, and his students were free to choose and change their work space as required. Several groups were collecting, weighing, categorising and comparing the litter in the courtyard bins with that which had been dropped on the ground during the previous lunch break, while others were collating and analysing their observations of lunchtime litter-dropping behaviour. The students compared that days' findings with the findings from previous days and earlier made predictions. There was a general air of excitement and activity, as the students passionately discussed and negotiated aspects of their tasks, sought assistance from other groups as required, and freely reported their findings, advice and ideas to each other. Periodically the students would seek assistance from David as they represented their findings in the form of graphs and tables.

Towards the end of the afternoon, David called his students into the classroom to present their results, share their analyses and discuss their ideas. He facilitated a class discussion in order to model how to use evidence to justify ideas as the students developed suggestions for reducing the rubbish and for changing littering behaviours. The students were encouraged to identify possible human–human and human–environment relationships to explain their data. The ideas developed during the class became the basis for the students' decision to report their findings and suggestions to the school council. The students listened attentively to each other and eagerly participated in the class discussion. Although David occasionally posed questions in order to assist the students to progress their thinking and extend the analysis of their results, the students were self-motivated, highly engaged, and clearly owned and directed their learning and the generation of new ideas. This project exemplified David's belief that effective education teaches students how to question their knowledge, and should be measured by "behaviour change" facilitated by that new knowledge. David's pedagogical approach to this lesson is represented in Fig. 5.3.

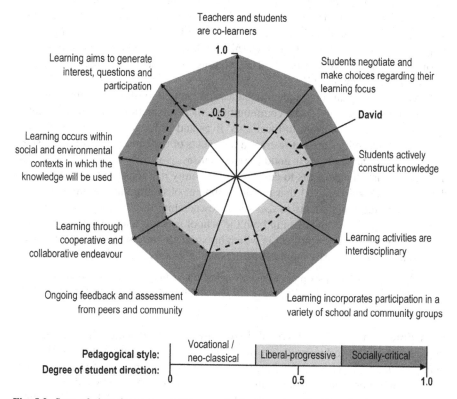

Fig. 5.3 State of play: the pedagogical approach of one teacher, David, to implementing SSP at Ocean Primary School

David explained that his aim in developing these types of learning activities was not just to "engage the kids" but to provide "real life learning", that is, an:

> authentic learning platform…a hat to hang stuff on…in other words, rather than just say write a story or think of something you make up, we have a lot more in common experiences that we can all write about.

He noted that he also explained the importance of authentic learning to his students:

> I say to these kids…I've taught you these tactics…how to learn spelling, how to do your tables, how to do subtraction the right way…but it's no good you being able to say oh you just do this, you've got to be able to say why you're doing it and why it's important…and it's the same with the environmental work.

Understanding Pedagogy David had clear and strong beliefs about appropriate and effective pedagogy for primary school classrooms, particularly in relation to the learning outcomes for SSP.

David described teacher-directed pedagogy as "probably conscientious" but a failure in terms of allowing "the kids to have some input" or accommodating "different learning styles". He explained that because "the teacher's sort of orchestrating and conducting more, it's not collaborative" and therefore "it seems not to be as empowering to kids, or of authentic learning". David believed that any activity that involved "collaborative learning and negotiating" provided unequalled opportunities for "fantastic learning".

David was adamant that authentic education came only through interactions in and around the students' local communities. Such interactions provided learning opportunities most "relevant to local kids' lives and experience". David explained that such opportunities are not restricted to aspects of the natural environment, stating that: "involving parents is good [be]cause it involves the wider school community" and assists the students to develop "community relations with people outside the school". David believed that the development of community relationships, as well as enabling the students to feel some "ownership of what they are doing" at school would reduce vandalism in both the school grounds and surrounding public areas. He described a new project in which his students were working with the local council to decorate nearby public structures with "murals and mosaics":

> they're all year [Grade] six projects and it's going to hopefully address that sort of thing [vandalism] so the kids don't come back and burn down your school or your bins or [kill] your chooks.

Impediments to Socially-Critical Pedagogy David attributed his success in implementing SSP to having "a lot of support" and the opportunity to work with many passionate teachers. However, he believed that the process of change would not continue, nor would changes already accomplished be maintained, without a constant, ongoing effort from himself as the coordinator. He noted that he attempted to prevent such a collapse by "spreading the load". Rather than establishing a single "environmental club which will then collapse if I go" he explained the strategy of having "different teachers take on different aspects" of student involvement with

SSP: "I've got teachers in my team that will lead certain areas...so there is a garden club, a compost club, there's a sea savers". David agreed that implementing the socially-critical pedagogy demanded by SSP would be easier if:

> in that perfect world, more of your staff thought like you did, but that'd be boring, they shouldn't all think the same but maybe have the same end or ideal...everyone has different priorities and things.

David believed that the outdoor facilities at Ocean Primary School had been an important factor in effectively implementing certain aspects of SSP, particularly the school's ability to "provide a nice learning environment...a nice place, we go outdoors, we can have our outdoor learning environment here". He explained that before the development of the outdoor areas:

> because we're far out we probably do less excursions than the inner city schools, you know it's a big deal to get the transport for a grade to go to the museum or Scienceworks[2] or something like that...we try to...but you spend a lot of time in travelling in and then you quickly do your thing and come back.

However, unexpected issues had arisen from the desire to develop these outdoor areas, particularly in terms of the lengthy process required to obtain even the most basic resources due to the different values and opinions of the parents and the community. He described the "pedantic" decision-making process of the school council regarding issues such as what garden mulch or garden soil to use, or where to source chickens for the farm as a result of "political correctness and occupational health and safety where they don't want to be seen to be liable down the track". David explained that "I understand that in a way, but it's got to the stage where it [public liability] has become a fear" and "can be frustrating" when trying to initiate simple changes in the school.

Similarly, David indicated that "money" was certainly "an issue", stating that: "I'm sorry to be so fiscal about it, but it's true, you know these days with global [whole school] budgets...money needs to be divided up, and there are other authentic things [for example] the reading books are getting tatty". David suggested that the school could more effectively target spending, for example, "when we get the reading books, let's look at the themes in the books".

David also believed that money could be used to increase community engagement with students by providing "a carrot...to buy people in". He saw this as one part of the solution to the difficulty in getting parents and others in the local community to commit their time to school projects. He noted that it was always the same few parents who attended working bees at the school, and who assisted with the construction of the outdoor learning areas used by his students.

He explained:

> I find this area to be in the comfort zone a lot, in other words, if you have a look around here there's nice houses and that which is great, really nice parents, they really are, don't get me wrong, but, we had a meeting last year [to begin implementing SSP] and we had so few attendees...we were trying to sort of explain what we were about and what was happening and why it was important to pass these things on to our kids...it was very poorly attended unfortunately.

---

[2] Scienceworks is an interactive museum of science and technology in Melbourne, Australia.

Despite the lack of enthusiasm from parents, David stated that "I still like the principle of having community meetings or involving parents and we try to stick things in local papers and have community involvement that way".

David believed that the ultimate effectiveness or success of SSP equated to "you can tell me the right answer if you know it, but does it translate into action?". He agreed that "some of those changes you can't measure".

5.5   Sirius College

Sirius College was a large independent school that catered for both primary and secondary school students across several campuses. One campus, a co-educational primary school which catered for almost 500 students, was located approximately 20 km from Melbourne, in a residential middle class suburb with a local population of over 20,000. This campus had been designed to fit in with its natural bush setting. The grounds were extremely well-maintained, and incorporated both unstructured open spaces and well-equipped play grounds interspersed with native gardens and untouched native bush. The campus was within walking distance of a range of urban and natural landscapes including a busy shopping district, a parkland reserve and a local creek.

One teacher, Cathy, shared her experience of working to establish SSP at Sirius College, and her efforts to incorporate SSP ideals into her classroom teaching.

5.5.1   Cathy

Cathy had been a generalist primary school teacher for 30 years, the previous 11 of which had been at Sirius College. In addition to teaching Grade 4, Cathy was the middle school coordinator and was responsible for the implementation of SSP. Despite extensive experience teaching all aspects of the primary school curriculum, including science, she described her previous experience in environmental education as "None!" stating that "I'd do the odd unit here and there, always have… but this [SSP] is a totally different way of looking at it".

SSP Implementation Sirius College had been a SSP five-star accredited school for 12 months. Cathy attributed "part of our success" in implementing SSP so effectively to the inclusive way in which the program was developed throughout the school, stating that "everyone supported it really well" and most importantly, "everybody had a hand in it", such that "even though it's been a lot of work for the person who was in charge of the enviro[ment] program, everybody's been there to back them up".

Cathy noted that this level of support reflected the fact that both the teachers and the school "management team...have always been...interested in developing a program [which was] environmentally...focused" because "we have a beautiful [bush] environment" at the campus. This made the decision to implement the SSP easy, because "the school as a whole...just thought that it was something that we really needed to do". Cathy believed that the SSP focus on providing new perspectives for "looking at our natural resources and seeing ways in which we can use them better and make less of an impact on our environment" provided a way in which to develop a unified approach across all levels of the structural hierarchy of the school.

Most importantly, Cathy explained that Sirius College embraced the notion that pastoral care was central to the development of a school community which most effectively supports all students to do their best. This vision was not limited to the students, and was exemplified by Cathy's description of her work environment: "we had a head of school that was fairly progressive in that she believed that [the teachers] could try new things". This attitude was "part of the [school] vision" which encouraged both the teachers and the students to "have a go". Cathy explained that the teachers at Sirius College had "always been encouraged and given great opportunities to do PD [professional development]". The teachers understood that if they "came up with something...if it [was] reasonable [they could] have a go at it". Furthermore, it was considered "okay" for a teacher's idea to "fail" because undoubtedly they would have "learnt something out of it". Cathy believed that these attitudes meant that the teachers at Sirius College were comfortable with agreeing to implement and trial a new program such as SSP.

The SSP implementation process at Sirius College began with professional development sessions which aimed to raise the teachers' awareness of current environmental concerns and which therefore highlighted the role of such a program. The teachers explored human–environment relationships by calculating their own eco-footprint. Cathy reported that this activity:

> had a huge impact on the staff as a whole and I think that has been the secret to getting to where we have so quickly...everybody...has seen the benefit of such a program...I think it made a huge impact when we did the...eco-footprint on each of the staff members.

Cathy explained that this had been a very important personal motivator, and stated that it had changed "my life as a teacher...and my family life" as:

> I look at the world in a totally different way...turning off lights and not having as long a shower...it really does make you so much more aware...it affects your whole life...I'm much more aware of environmental issues now than what I was before I did the program...I used to hear it, yes, I was concerned, but I wasn't [at] the forefront of doing something with it. I am now, and I think that's made a huge impact on my life personally.

Cathy indicated that Sirius College harnessed the momentum for change initiated during the professional development sessions by ensuring that the teachers owned the ways in which their new understandings were incorporated into their professional lives. She stated that all of the teachers contributed to a "big brainstorm about what our ideal school" could be, and that "it was from that that we...did our overall vision...our enviro[ment] policy". This process provided opportunities for the teachers to identify the most effective manner in which to implement SSP ideals throughout the school, and in Cathy's case, ways in which to most effectively involve her Grade 4 students.

Every year at Sirius College, each class participated in a leadership skills program by undertaking a special responsibility. As plans for implementing SSP were being developed, Cathy had been concerned that her "year [Grade] 4s were sort of left out there on a limb" as a worthwhile special responsibility had not yet been identified. During the brainstorming process Cathy identified SSP as a useful vehicle for developing aspects of leadership and social responsibility; assisting the students in "being a more responsible person about their environment…thinking about the issues which are so important to society at the moment with water, and energy use". This prompted her to focus on learning to "care for the environment" through implementing SSP as an environmental leadership role for her class.

Understanding SSP Cathy described SSP as a program which offered a great variety "of things that you can intertwine with your curriculum program" and which provided opportunities for participation which assists students to "see that taking action is so important, and just little bits make such a big difference". Cathy's understanding of the role of SSP was best represented by her comment that it provided students with opportunities to develop "not only an idea of basic knowledge about the world, but [also] how they can make a difference". She stressed that personal development, in terms of ability to take action, was an important aim of SSP, stating that:

> I think that they [the students] can actually see that they can make a bit of a difference, even one single person can make a bit of a difference…It's that sort of awareness of things that they [the students] can actually do themselves, and not only help their communities, but also help the environment.

She illustrated her understanding of SSP through examples of the ways in which she encouraged her students to be actively involved in a wide variety of projects. For example, her students worked with the local council on a tree planting program:

> We have our tree planting day where the year [Grade] 4s actually organise it all. They supervise here on campus…pre-school up to year [Grade] 2 to actually plant [indigenous plants] on campus…[grades] 3 to 6 actually work with…people from council and plant out on [a local] reserve.

Cathy also described the results of a local water testing project:

> we go out and do the Waterwatch[3]…test the water and…we found…I think it was phosphorous…in higher amounts than it should have been, so they [the students] discussed…what should happen and so the…person who led this…from Waterwatch, actually rang up the EPA [Environment Protection Authority][4] and told them this.

---

[3] A community water quality monitoring network headed by the Australian Government Department of the Environment, Water, Heritage and the Arts.

[4] A statutory authority.

Cathy described the SSP as "a huge success", not only in terms of student engagement, "because the children have really taken to it", but also in terms of the behavioural changes and personal development she had observed. She stated that "it's amazing" that her students would leave "the classroom for a specialist session and in the middle of winter, they'll turn off the lights on you". She also noted that anecdotal evidence suggested that her students had "introduced lots of the things that we've introduced at school at home. So it's [SSP] making an impact on the various home lives as well".

SSP in Practice Cathy found it difficult to identify a single lesson to best portray her approach to implementing SSP. She considered her role as SSP coordinator most important for developing extra-curricula school-wide activities, as in her own classroom she incorporated sustainable school ideals in all classes. She nominated a session in which her students were continuing in their implementation of a 'no-rubbish' lunch policy.

Cathy described her classroom as "vaguely organised chaos". The sounds of student–student chatter and activity could be heard well before entering the classroom. Groups of students and the work in progress had spilled well beyond the classroom door, along the adjacent corridor and into nearby outdoor spaces. The scene from the classroom door was one of disordered furniture, scattered writing materials and free roaming pets, around which groups of students were actively engaged. It was clear that the apparent pandemonium was actually a dynamic learning space that the students continuously modified to facilitate their requirements for interaction and learning.

The students were enthusiastically implementing a 'no rubbish' policy for lunches across the entire school—an SSP accreditation requirement introduced as an idea by Cathy, but designed and managed entirely by her students. The students were assessing the effectiveness of their initial advertising campaign (which had incorporated the creation of the 'Nude Food Dude' comical character) by collecting classroom rubbish bins, weighing, counting and classifying rubbish. Groups of students were working to present their findings in order to inform classes of their progress towards the nude food ideals. The students needed to determine how to best collect, represent, display and explain their data in a manner that portrayed their findings to other classes. Cathy supported student collaboration and peer learning by acting as a sounding board and advisor, and providing direct instruction only when necessary. At the end of the session the students came together for a class discussion. Cathy contributed only when necessary to model appropriate questions, ideas or responses, or to assist the process to continue. Each group of students shared their findings, provided critical feedback to others, and were encouraged to accept feedback with a degree of critical consideration. Cathy's pedagogical approach to this lesson is represented in Fig. 5.4.

Cathy noted that many other SSP-related activities had been initiated by her students. One student's request to participate in a bird nesting program resulted in the students successfully applying for a grant and conducting research into birds' nesting needs. Cathy sought assistance from wildlife experts who led a field trip for the students, and helped them to construct appropriate nesting boxes that were placed in

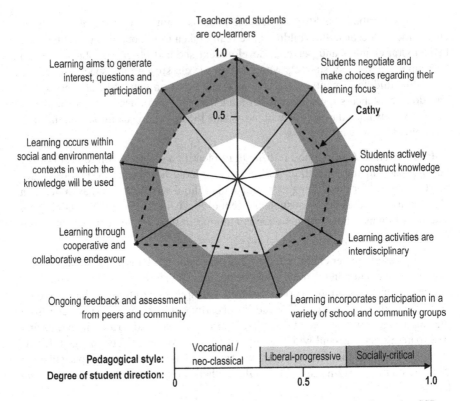

Fig. 5.4 State of play: the pedagogical approach of one teacher, Cathy, to implementing SSP at Sirius College

trees around the school. The students continued to monitor the nests, collecting data that contributed to other literacy and mathematical projects. This had not been the only time Cathy had sought assistance from people with specialised expertise. She described her experience of a program in which the students designed and built a model vehicle powered by a hydrogen fuel cell. The students experimented with different materials and different designs in order to maximise the speed of their vehicle. Cathy invited a secondary high school teacher to assist her students to answer their own questions. This was so successful that the learning and engagement opportunities for her students surpassed the possibilities of other activities such that Cathy extended the initial 2 week program to an entire term. She commented that this had "been an exciting journey" as her students had taught her about the properties of a periodic chart and basic chemical concepts related to hydrogen fuel cell technology, and that learning from her students in this manner was what "makes coming to school enjoyable".

Understanding Pedagogy Cathy's classroom practices reflected the strong convictions she held regarding the educational validity and role of different pedagogies. She described traditional vocational/neo-classical pedagogies in science as being "too prescriptive". She rejected the notion represented in the hypothetical scenario that portrayed a vocational/neo-classical pedagogy (Table 4.3) that teachers must "maximise the chances of successful completion" of a curriculum activity, and referred to her own experience in giving her students the space and freedom to learn from their own mistakes:

> I think one thing that we got out of that [hydrogen fuel cell project[5]] …was the mistakes that the kids made…I mean, yeah occasionally they were disappointed about it, but the learning that went on from the mistakes…to see the way that they attacked that and the testing and investigating that they did…I mean there are times when you want them to learn particular skills, then you teach them that skill, but if it's just something that you want them to learn about a particular topic, I think they have to make those mistakes.

Cathy considered it important "to consult the kids about where they want to go" because "they come up with things that you may not have thought of and I think that that's what's exciting". She detailed her need to embrace socially-critical pedagogies for their ability to create "exciting" opportunities in the following way:

> I'd stagnate if I had to do the same thing over and over again…I mean we have to be motivated to get the kids motivated…if we're not really excited about doing something, how can we make the kids excited about doing it, and I can't see that you can get excited about something that you've done twenty times before.

Impediments to Socially-Critical Pedagogy Cathy did acknowledge however, that implementing a socially-critical pedagogy, or any pedagogy other than a vocational/neo-classical approach, was not always easy. She believed that during her "earlier days of teaching life" she may have felt the need for "a little more control" in the classroom than a socially-critical pedagogy provided. She understood that implementing socially-critical pedagogies requires practice, but noted that "part of getting that experience" requires teachers to accept that "there's lots to be learnt on your part [as the teacher] as well, you don't know everything" but that when "you throw yourself in there, you learn so much more".

Cathy considered the "beauty" of the SSP to be that it "doesn't get mundane" and constantly challenges teachers because "there's something new all the time". She illustrated this with the rhetorical question "We went through the initial part of setting up the Sustainable Schools Program, we got our five stars, [but now] how do we keep going?". In answer to this, Cathy identified the development of collaborative relationships with people and organisations outside the school to be most critical for the successful continuance of SSP. She believed that this requirement was the most difficult component of her work, and one which required a significant investment in "time":

> I think that's probably the biggest impact on what we do, not having that time to sit down… make our contacts…cruise on the web…time is definitely the killer…I just feel that things have changed, there are so many more demands on us now than what there were when I first started teaching…there's just so many things you have to cram into your day…your emotional problems with kids, and your emotional problems with parents…in a more perfect world I'd have more time.

---

[5] A project that required students to design, build and race a model vehicle powered by a kit hydrogen fuel cell motor.

5.6   South Bay Primary School

South Bay Primary School was a small co-educational government school attended by 100 students. It was situated in a small, semi-rural region approximately 50 km from Melbourne, with a local population of about 375. The local community was composed predominantly of pastoral families, many of which had been financially devastated by severe drought. A small, but growing portion of the community was represented by hobby farmers whose main employment was outside the town.

The school was preparing to be mostly re-built during the upcoming summer holiday period. The new school was designed to model best practice sustainable development principles, particularly in relation to minimising future energy requirements for lighting, heating and cooling. Similarly, the new classrooms had been designed to enable best practice pedagogy by providing a variety of indoor and outdoor learning spaces. The school was within walking distance of a variety of natural landscapes, including a major river, native forests and grasslands, fauna and flora reserves, and rocky hills.

The principal, Helen, and the SSP coordinator, Andrew, shared their experience of establishing SSP at South Bay Primary School. Lisa, the school's science teacher, shared her experience of working to incorporate SSP ideals in her classroom teaching.

Implementing SSP   Helen, the principal of South Bay Primary School, had identified SSP as a potentially important component of teaching and learning, and had overseen the first 9 months of its implementation. Helen described SSP as being much "more than environmental education" and more like a future-oriented education that aims to provide society with the skills to sustain human–environment relationships into the future. At a personal level, she viewed the implementation of SSP to be "about sustaining [her] own life" by enabling her to actively incorporate her environmental philosophy into her education role. As the school principal, she identified the program as "a vehicle for whole-school change" through which she could begin to transform entrenched and outdated teaching practices by developing "a new way of teaching with a new way of learning". Helen believed that effective pedagogical change would occur only if the teachers took "ownership" of the change process, and that in order to do this, they needed to develop a shared discourse through which to explore, share and develop new ideas. She supported this by enabling the teachers to attend professional training sessions at CERES, and setting aside time for collaborative planning and curriculum design. She encouraged personal learning by requiring the teachers to maintain reflective journals as part of implementing SSP.

Helen had allocated the day-to-day responsibility for coordinating the implementation of SSP to Andrew. Andrew had been a qualified teacher for just 3 years. He had spent this time teaching students of various ages at South Bay Primary School, and had accepted many additional responsibilities, included student health, critical thinking, and science and technology education. Andrew was managing the school's transition to a new state government curriculum and was contributing to the planning process for the building of the new school.

Andrew's personal interest in SSP was focused on the opportunity to develop their "new school around sustainability". Andrew viewed SSP as future-oriented education that aimed to protect human life by developing the students' knowledge "about the environment" as well as the "effects" or future "consequences," of human activities on the environment. He was adamant however, that he would not allow the implementation of these educational goals to jeopardise the introduction of a new curriculum at South Bay Primary School. Instead, the initial SSP modules were to be implemented as discrete sustainability topics, each a context through which "other curriculum outcomes" could be achieved. Each module would be facilitated by one teacher. Andrew did not allow his lessons to be observed, stating that he taught "nothing of interest" because the current SSP topic was the responsibility of the science teacher (grades 3–6), Lisa.

Andrew believed that in conjunction with a newly designed school, SSP would provide opportunities for students to "take ownership of how they affect the environment" and to develop their understanding that achieving environmental sustainability is "a collective process" in which every person's actions matter. However, he indicated that the students at South Bay Primary School had not, nor should they have, any role in the implementation of SSP, because student leadership and choice was problematic—the students "don't have a grasp of what they're doing" which meant that "learning outcomes might not be met". He believed that increased financial resources would assist in the implementation of SSP, especially through the provision of expert tuition, which he believed to be highly beneficial, and which could be offered in the form of community participation where people attended classes to "talk to the children" or through excursions to established environmental educational centers.

5.6.1   Lisa

Lisa had been a qualified teacher for just 3 years. She had spent this time as a generalist Grade 5–6 teacher at South Bay Primary School. Because Lisa was the only teacher at the school with university science training, she was responsible for teaching science and SSP to all of the students in grades 3–6, in addition to coordinating the delivery of the science and SOSE curricula, and leading the transition to a new curriculum in these subjects. Lisa noted that in addition to teaching the environmental understandings required by specific learning topics within the science and SOSE curricula, she voluntarily coordinated groups of students to participate in special environmental activities, such as Clean up Australia Day[6] and local community tree planting activities.

Understanding SSP Lisa strongly supported both her principal's decision to implement SSP, and Andrew's efforts to assist this to occur. She considered herself to have "always been into the environment…and recycling" and believed that her "personal passion" would assist her "to stay interested and focused in making sure

---

[6]A community rubbish-collection event organised by the Clean Up Australia not-for-profit conservation organisation.

it happens" and that the program would become fully implemented across the school. She believed that her personal interest would help her to make the most of the new educational opportunities SSP would present to her, because she believed that as a teacher she had "the power to get out there to kids" and therefore "able to put it [SSP] into the school".

Lisa considered the main role of SSP to be about "educating the children about the future" by "teaching them good habits so that life in the world becomes sustainable". She saw SSP as a vehicle through which to establish such habits as accepted and unquestioned routines where "we turn things off, we do things the right way, we don't have to be pushed to do it or asked to do it, it just happens—it's just a natural thing that you do". However, as effective future-oriented education, Lisa noted that SSP must incorporate two essential elements: (i) opportunities to change students' attitudes towards their consumer practices; and (ii) opportunities to encourage students to take responsibility for their own actions and to speak out about the actions of others.

Lisa believed that society would only become more environmentally sustainable when students are taught how to embrace attitudes that reject "all that consumerism" so that they begin to understand that they "don't need everything brand new", and that they don't need to "have every different game boy, play station or whatever it is going around". She understood however, that this type of learning would only contribute to "long lasting" change if students also developed "some sense of responsibility" and "the passion to go away and learn" for themselves what they must be "putting into place at home and around the community". Lisa believed that her role as a teacher in achieving these educational and SSP aims was to develop her students' interest in the world, and that in science, this required incorporating a hands-on approach to learning, stating that "we do the hands-on experiments because it really draws them [the students] in".

SSP in Practice Lisa chose the second of a two-lesson unit about water to highlight her preferred method for incorporating SSP ideals into her teaching. She explained that this science unit would be taught to all of the students in grades 3–6.

Lisa's classroom was immaculate. Although it was towards the end of the day, the carpeted floor and table tops were perfectly clean, and there was an air of well-maintained order. Four long rows of neatly aligned desks ensured that all of the students faced the prominent whiteboard at the front of the room. The classroom walls displayed the students' work pinned in neat rows, while all learning materials and books were stacked neatly on the side benches or packed away in plastic containers. Between the board and the front row of desks was a low table upon which sat a single set of kitchen scales and a plate of desiccated fruit. As the students entered the room they were instructed to find their seat, sit quietly, and to place their books neatly on the desk in front of them.

Lisa opened the lesson by reviewing the progress of their science investigation, reminding the students that they had seen her cut various types of fruit into pieces, weigh them (she pointed to weights recorded in a table on the whiteboard) and then place them on a tray so that they could be left outside in the sunshine. She thanked the individual students who had accepted responsibility for collecting the tray at the end of each day and placing it outside each morning. During this introduction the students answered specific questions posed by Lisa regarding what they had noticed about the fruit throughout the week.

Lisa continued the lesson by re-weighing each piece of desiccated fruit. Individual students were chosen to read the scales, but were instructed to not touch the fruit or the scales. All hands eagerly went up for the chance to be chosen, and although this involved no substantive hands-on activity, those who read the scales were clearly thrilled to contribute to the lesson in this manner. Lisa corrected any errors in the scale readings, and recorded each weight in the appropriate column of the table on the whiteboard. Lisa assisted the students to calculate the percentage of water loss for each piece of fruit by dictating the calculator process to ensure that all of the students obtained the correct answer as it was written on the board. She instructed her students to copy the table into their science books and to write one or two sentences to explain which fruit has lost the greatest amount of water. The students were reminded to do this without talking. Throughout this lesson there was no class discussion, nor any indication that either the teacher, or the students understood the purpose of this project. Despite this, the students were compliant and seemed to enjoy the lesson. Lisa's pedagogical approach to this lesson is represented in Fig. 5.5.

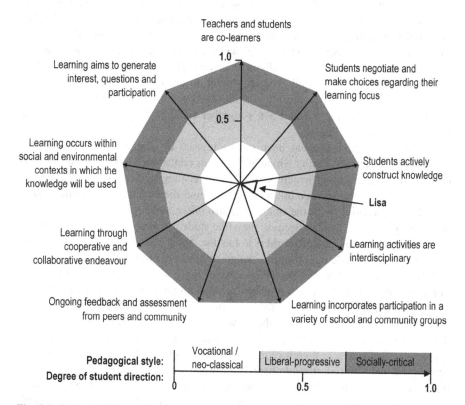

Fig. 5.5 State of play: the pedagogical approach of one teacher, Lisa, to implementing SSP at South Bay Primary School

Understanding Pedagogy Lisa had an excellent understanding of the benefits, differences and similarities of various types of pedagogies, particularly in relation to the application of teacher-directed versus student-directed learning, and the importance of "student involvement", suggesting that "if they [the students] own it they have a different attitude towards it". When discussing the hypothetical scenarios, she commented that:

> the first one [liberal-progressive pedagogy] is sort of teacher-focused…no student negotiation in it, whereas the second one's [socially-critical pedagogy] got some student involvement…they have power over it, and if they own it they have a different attitude towards it.

Lisa described the implementation of SSP at South Bay Primary School as essentially "teacher-led" and not yet at the stage where "it's students making the decisions on what they're doing". She commented however, that the teachers at the school "certainly were getting out there and starting to get involved in it" and that "hopefully one day we'll get there".

She also noted her personal desire to change her pedagogical approach, stating that "I'd love to be at [hypothetical scenario] number two [socially-critical pedagogy]…hopefully one day we'll get there". Lisa's desire to implement a more socially-critical pedagogy stemmed from her observation of improved student engagement and learning during several specific activities that didn't adhere to the strict teacher-directed focus of a traditional vocational/neo-classical approach. She described the school "science and maths night", based on environmental themes, for which the students "decided what experiments and what activities they want[ed] to show their parents". Lisa also recalled the success of a recent project in which her students worked in multi-aged groups to paint an environmental mural at a local public site where "the kids are all in multi-aged groups to go on the excursion, instead of putting all the [Grade] 5–6's together we split the grades, and the 5–6's are going with a Prep grade".

Lisa explained that she has been most enthused by a recent lesson along the banks of the local river, during which her students contributed to the organisation of the lesson by determining which parts of the river they wished to visit in order to answer some specific science questions. She found the benefits of combining student decision-making with the outdoor environment and hands-on learning to be really "inspiring as a teacher because they [the students] were so into it" and led her to the conclusion that the real world is in fact the "ideal classroom", and that school education must include opportunities "to go beyond your four walls".

Lisa commented that despite the success of these special activities, the school implementation of SSP had essentially excluded all students from the larger projects. Although she believed that the creation of a frog pond might have been in response to student interest, most things were not. She stated that, for example, "I don't really know where the veggie garden came from" because it "just seemed to appear one day" and was therefore most likely a project undertaken by parents and/ or the principal: "I think it might have been principal-led, or higher staff, or something like that". Lisa attributed this fragmented approach to SSP and the slow uptake of more socially-critical pedagogies as a reflection of the fact that the school community encompassed many "different people", some of whom "are more passionate about it [SSP changes] than others":

> you've got different people who are more passionate about it that others, so they'll make sure they're doing it, whereas others just…it doesn't matter now…sort of thing. But ideally it would be good to see it all happening that way [socially-critical pedagogy], but I don't think that any time soon it will be.

Impediments to Socially-Critical Pedagogy   Lisa identified teaching space as the most significant impediment to her ability to more fully embrace the pedagogical changes inspired by SSP. Lisa considered the lack of a suitable "area where you could work outdoors" as particularly limiting, because she did not feel that she could ask her students to sit "outside for an hour and a half under the direct sun". She looked forward to the completion of the new school buildings, designed with environmental ideals in mind, and in which "every classroom has access straight out into outdoor learning areas".

Lisa suggested that the implementation of more socially-critical education requires a different style of school building in which it is easy to move students between different types of areas without "constantly moving rooms around" or "moving furniture around to accommodate" opportunities for shared learning through interaction between classes. Lisa was obviously proud of the fact that the soon to be built school had been designed with environmental ideals in mind, and indicated that the structure of the new school would definitely assist with the implementation of SSP. However, Lisa also admitted that although she might complain about a lack of time or physical resources, these were "really just excuses" and that her work would benefit more from "just re-organising the way things are structured or getting rid of things that aren't needed".

5.7   West Quay Primary School

West Quay Primary School was a small co-educational government school attended by 110 students. It was situated in a small rural community about 100 km from Melbourne, with a local population of about 1,000. A nearby manufacturing and industrial centre enabled families to seek work outside the town in order to supplement their declining incomes from drought affected agricultural activities. Plans to up-grade the school grounds as part of the SSP implementation had also been postponed due to the water shortage. The school aimed to support students on their journey to becoming active participants in society.

The principal, Philip, and two classroom teachers, Fran and Simon, shared their experience of working to implement SSP at West Quay Primary School.

Implementing SSP   Philip, the principal of West Quay Primary School, identified the implementation of SSP as a school "goal" to accompany planned improvements to the school grounds. Philip had used the first year of the implementation of SSP as a vehicle for improving the operational practicalities of the school in order to establish a working model of sustainable living that would be financially beneficial. Simon agreed, and identified the financial aspects of SSP as one of the most important benefits of implementing the program, stating that "there's a lot of money involved…a lot of money poured into this effort and it gets used right in the schools". Simon explained that "we've actually purchased quite a few things…in the name of sustainability", including for example "remote control boats that we raced in Sydney" and items for the school such as "solar panels [and] a worm bin". However, Philip believed that achieving the full range of educational outcomes from SSP did not rest with securing additional money or additional resources, but most importantly required improving student engagement and learning through pedagogical change. He indicated that the overriding motivation for implementing the program was his passion for "connectedness with the community", an essential foundation in the development of student understanding of human–human and human–environment relationships. He believed that this would facilitate the production of environmentally and socially "responsible, ethical individuals".

Philip stated that initially, most of the teachers "didn't want to be involved" with SSP. He noted that he had needed "to lean on some people" and "encourage others". Fran, for example, described her participation in the implementation of SSP as being "driven" by Philip. Similarly, Simon had not volunteered to participate in SSP. He stated that "I probably didn't start out that interested" but "the principal chose a group of teachers to get involved". He explained that he had been chosen because "my role is science and that fits in quite well with it [SSP]". Simon also noted that Philip's ongoing interest in SSP had been essential, because the program would not "have got very far without the school making it a definite goal".

Philip described the facilitation of change as a difficult process of "getting people on board" and challenging long held attitudes and beliefs. Fran agreed, and felt that the introduction of SSP into West Quay Primary School had faced many difficulties. She described the early stages of SSP implementation as somewhat hurried and at times, seemingly directionless which she attributed to two main issues: the lack of motivation of some teachers to participate in the change process, and the challenges faced by the teachers who wanted to participate but were unsure how to do so. Fran observed that many of the teachers at West Quay Primary School were poorly motivated simply because they did not share the environmental ideals of SSP. She explained their reluctance to engage with the initial 'reduce, re-use, recycle' activities as "they're of a generation when there was heaps of water, and this is how you use water, you did water your lawn all the time". She commented that, even in the current drought conditions, such teachers held the attitude that "I pay water rates therefore I have the right to water my lawn". At school she would "hear people [the teachers] saying you can cheat in the back garden with the water restrictions" and would "think why? Who are you? Who are you cheating?".

Philip explained that the most difficult aspect of the early stages of implementing SSP was the school-wide change towards "a more child-centred curriculum" through the development of "inquiry approaches and integration of thinking curriculum". He believed that the lack of teacher motivation to engage with this "curriculum development" potential of SSP was related to the teachers' unwillingness to question the pervasive culture of belief that he described as "I am the font of all knowledge and I spew forth". He believed that the entrenched vocational/neoclassical teaching practices at the school were supported by a routine of "plan[ning] to the nth degree", such that the teachers had become "extremely inflexible" and fearful of change, and had lost touch with the ultimate goals of education. Fran explained that the apparent inability of some of the teachers to enact these changes was not simply a form of deliberate "resistance", but rather:

> it's just that people can't imagine how it's going to work in their classrooms, what they're going to do, so some people are confused about what they can do, and some people are scared of changing, throwing everything out, or even fifty percent out, and trying a different way.

Both Philip and Fran commented that many of the teachers considered SSP to represent a significant amount of extra work to add to the current curriculum rather than a new approach to their current practices. Fran suggested that even for some of the teachers who agreed that "it [SSP] had to be a whole school thing…[and] it had to be built into the curriculum" this was actually "really challenging" to achieve:

> it's especially challenging for teachers who have taught in the same classroom in the same way for twelve years or so…to actually get their head around the fact that…it's not an add on…it's part of the curriculum, it doesn't have to happen outside school, it happens in school, it is part of your day with children.

Philip explained that although some of the teachers had been "late subscribers", agreeing to participate only after observing the successful efforts of others, other teachers had remained steadfastly unwilling to trial new pedagogies. Fran explained that this inability, or unwillingness, of certain teachers to engage with any part of the program was one of the most difficult aspects of the implementation of SSP at West Quay Primary School. She stated "if you just have your little grade doing your block of things, that's divisive, it's not part of [the school] community". She explained that "I think that's something some teachers guard pretty strongly as well, they guard their own class…my class does this [and] whatever you do is up to you…you do see that in teaching".

Fran reported that many of the teachers at West Quay Primary School willingly contributed to, and participated in, the implementation of SSP and "really did want to make changes". However, Fran explained that these teachers still experienced difficulties with the implementation process:

> at first we were quite lost, we didn't really know sustainability, and we felt that we had to get it moving so we went into a bit of a frenzy which didn't really solve too many problems and we floated around.

Philip stated that the support of the CERES SSP facilitators had been instrumental in developing the motivation and momentum for change by providing opportunities for the teachers to take ownership of the change process as "part of a learning community". Simon acknowledged that SSP provided excellent "support with really good professional knowledgeable passionate people...people who actually come to the school and talk to you, and support you". He noted that this support was significant, because "it's not like a lot of things [previous programs] where you're told to implement it and then either given a whole bunch of paperwork or given nothing, and then the thing fizzles out".

Despite achieving significant change however, Philip acknowledged that the gap between the rhetoric of SSP and teaching practices remained large. He strongly believed the development of new learning spaces within the school grounds would help to limit such rhetoric–reality gaps, although he conceded that many of these, particularly the planting of kitchen and native gardens, were on hold due to severe drought and associated water restrictions. Despite the delay in the development of outdoor learning areas, West Quay Primary School had completed the mandatory SSP energy, water and waste audits, and the establishment of a small, temporary vegetable garden had enabled the students to grow vegetables from seed. The students sold some of their vegetables to parents at the school gate. Despite Philips' vision of the school as an integrated community in which learning can occur, he reported that the students at West Quay Primary School had not been involved with the design or construction of the new outdoor SSP learning areas, although they had been consulted about what they most wanted from an outdoor play area.

5.7.1   Fran

Fran had been a generalist Grade 4–5 teacher for 7 years, the last 6 of which were at West Quay Primary School. Fran's additional responsibilities included LOTE and Occupational Health and Safety (OHS).

Understanding SSP Fran stated that although her participation in SSP was initially "driven" by Philip, she acknowledged always having had "an interest in the environment" which meant that "I was happy to put my hand up and say yep, I'm interested in sustainability". Despite this personal interest, Fran reported that the SSP professional development sessions had had a great impact on her way of thinking about the world. She explained that her understanding of 'environment' had broadened, because participating in such sessions "makes you start looking at *your* environment...the localised environment...in the buildings and so on". Above all else however, Fran noted that what had really stuck with her was the realisation that her students may grow up in a world in which it was not possible to travel to see, for example, "a gorilla in the wild". Fran was noticeably upset at the thought that her young students were facing an uncertain future caused by society's lack of action towards addressing current environmental issues.

Fran described SSP as a form of collaborative and socially transformative education which aimed to develop people's understanding that society has:

a responsibility to change the way we're currently operating for environmental reasons...to change the way we are currently living on the planet...to try to change the way we're currently using energy and so on...making changes that are favourable to the environment.

She explained that her views were not purely for environmental reasons, but that "it's for selfish reasons as well" as:

> we don't want to run out of water, that's the obvious [due to a current long lasting drought] one. We'd probably rather we didn't run out of coal and oil…but I think there are better forms of energy we could use anyway…so we want it in check for ourselves.

Similarly, she believed that such social transformation could be accomplished through implementing "changes that won't necessarily mean that our lives are worse off". She explained that this understanding was particularly important, as:

> some people think that they might be worse off, but they need to look at the big picture rather than the small picture, and realise that turning the tap off when you're cleaning your teeth is actually something that, if you'd learnt how to do that when you were 2 or 3 [years old], would not even be a factor in your life at all.

Fran believed that in order for environmentally responsible actions to become widespread "the first thing you have to do is change the mindset…it's a change in thinking". She viewed her involvement with SSP as an opportunity to help "this generation to be making the changes now into learning a different way of living… changes that the children can make now…will have long term positive effects…for the environment". She explained that:

> it's about trying to change habits and instil new habits that become the norm rather than the big something we have to think about…one of the teachers here I think hit the nail on the head when she said we need to make these responses automatic…so that [for example] when you leave a classroom, you automatically check that the lights are off.

Like many of the teachers implementing SSP, Fran believed that it was most important to introduce environmental and sustainability issue to her students in a positive manner in order to avoid creating feelings of anxiety and helplessness. She described her initial work to implement SSP ideals into her classroom as being focused on "waste and reducing, recycling and re-using". She explained that this seemed to be the most effective way to encourage her students to most easily and successfully adopt environmentally sustainable habits, while addressing authentic issues identified during the initial SSP school audit:

> some obvious things that we [West Quay Primary School] can address very early…some obvious waste problems…what I consider to be a waste of power, waste of energy in a big way…lights being left on…classroom doors are left open all winter with the heaters on… [and] not a lot of stuff being recycled.

SSP in Practice   Fran chose the first of a two-lesson unit she had designed to provide her students with opportunities to take environmental ideas "home" and "into the community" to highlight the manner in which she was incorporating SSP ideals in her teaching. The module incorporated two tasks, one of which required the students to use a computer. However, due to limited computing resources at West Quay Primary School, Fran's students had to work in two groups, and would rotate through the two tasks during consecutive lessons.

One group of students used computers to access a series of pictures of famous geographical landscapes, such as the Sahara Desert, Mount Everest and Uluru. Each student was asked to identify one picture that represented to them the "most pleasant place on Earth" and then answer a series of questions by selecting information from a list of facts. All of the information available to the students was contained within documents previously collated by Fran. Fran had not included images of local environments that the students were most likely to have visited and experienced for themselves. Although the students obviously enjoyed the opportunity to use the school computers, they seemed to have little interest in the task at hand, and finished as quickly as possible in order to chat about unrelated topics.

The other group of students worked to create a poster that they could use to deliver an environmental message to their parents. Fran explained that these posters would be displayed at a future parent–teacher discussion evening. Posters were to be a collage of recycled materials (mostly pictures from old newspapers and magazines) that portrayed a message about the importance to either 'save water' or to 'recycle'. The students seemed to work with little engagement or understanding of purpose, and very quickly produced posters with unimaginative environmental slogans (such as Save Water Now!) surrounded by an ad hoc collection of hurriedly cut-out images.

Throughout the lesson, Fran constantly asked the students to work quietly, concentrate and complete their work. She seemed to be either unconcerned or unaware that her students were completely unengaged with their tasks. Although the students did not significantly misbehave, they spent most of the lesson talking about unrelated topics, and seemed to ignore most of Fran's requests. Fran's pedagogical approach to this lesson is represented in Fig. 5.6.

Fran explained that her class also contributed to implementing SSP ideals throughout the school by managing an environmental reward system, initiated and designed by Fran to "slowly start changing the ethos" of the school. Her students audited the environmental behaviour of every class, monitoring actions taken to reduce the use of resources and to more effectively recycle rubbish. Points were awarded for actions such as turning off the classroom lights and heating when the room was not occupied, keeping the classroom door closed while the heater was in use, and keeping buckets under water taps in order to collect unused water. Each week, one class became the custodian of the "environmental frog" trophy. Fran admitted that this was nearly always won by her own class, and that, on occasions, she had directed that it be presented to another class as an incentive for them to participate more seriously.

Understanding Pedagogy Fran was confident in her understanding of what constitutes best-practice pedagogy for assisting students to develop the skills that they require to best shape their future contributions to society. She described a teacher-directed approach to learning as "a bit limited" and described the associated tendency to specify precisely what a student must learn from each particular lesson as a "switch-on-switch-off type of thing". Fran felt very strongly that if students were to become active participants and effective decision makers in their future society "they've got to have the guts to question" rather than "just swallowing everything they're told". She explained that she tells her students that one day "you'll be vot-

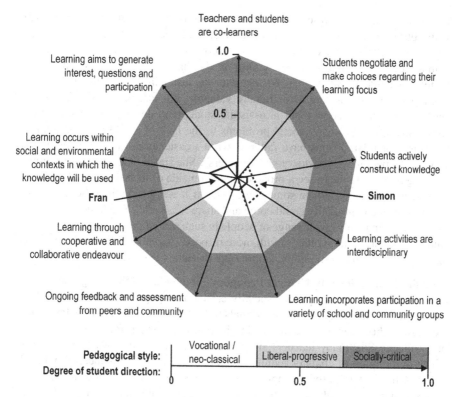

Fig. 5.6 State of play: the pedagogical approach of two teachers, Fran and Simon, to implementing SSP at West Quay Primary School

ing, you'll be making choices [so] you've got to be aware of what's [going] on, what's better choices…alternatives to our current use of power and so on". She also tells them to "question the newspapers" and "don't believe everything you read". Fran explained that she designed learning tasks to assist her students to learn how to assess the validity of information for themselves. She described a recent series of classroom conversations in which she had encouraged her students to think critically about newspaper reports warning of a potentially serious outbreak of bird flu:

> We've heard a lot about this [bird flu]…I'm sure it's a disease…we look[ed] very closely at the statistics that are involved…I think at the time five people in China had died and I'm thinking, you know, how is it that big?…people [are saying] oh that's terrible, five people, and I'm thinking how many people live in China?…so then we had to find out how many people live in China, and then, well, so five of them died, and I'm not trying to say that's not important, but I'm trying to get it into perspective…we need to get so many things into perspective.

Fran believed that students could best develop their questioning and critical thinking skills through the socially-critical pedagogical approach advocated by SSP, and that, most importantly, this approach recognised the importance of "community". She explained that SSP activities that encouraged multi-age collaboration provided "a realistic representation of the community" and therefore "a more worth-while purpose" to learning. She believed that unlike teacher-directed lessons, the authentic contexts of such student-directed activities provided "more possible out-comes". Fran also thought that such multi-aged collaboration should take place in a variety of contexts rather than be restricted to just within the school. She saw community-based learning as extremely valuable, not just in terms of achieving specific learning outcomes, but in assisting students to develop positive relation-ships. She attributed many social problems to the lack of a sense of community: "because our relationships have fallen down it's easier to not have a neighbourly way to like people we consider to be strangers". She stated that, as a part of imple-menting SSP, "I'd love to see a project happen outside of the school" and "I'd love to have a community garden somewhere, even if it was inside the school grounds but [where] the community could come and share it as well". She hoped that learn-ing and working with a wide range of people in such shared spaces would encourage students to learn to treat others "with more respect", and noted also that "you'd get less vandalism if the community can treasure or value the school".

Impediments to Socially-Critical Pedagogy   Fran believed that a significant bar-rier to the implementation of the socially-critical pedagogy required to teach SSP ideals was the time required for individuals to find ways in which to transform their well-established teaching practices. She believed that when faced with any new program like SSP many teachers simply required time to find ways in which to adjust to the types of changes required. She believed that "life is incredibly busy" and that the work required to implement change is often "put off until tomorrow". Fran stated that although "we all do that to some extent" it's important to remember that when stretched for time, people understandably "just do the same things…fall back on something that worked for them before". She described her role in the implementation of SSP at West Quay Primary School as requiring "enormous organisation…that takes away my time" but then admitted that "I think we could manage our time a lot better".

Fran believed that if SSP was to become more effectively implemented at West Quay Primary School "line[s] of communication to the community and to people who are like-interested" must be established. Fran explained that there were many local opportunities to establish links between her students and the community and appropriate local projects that the "school could get involved in", including for example, the ongoing management of indigenous and introduced plants along the banks of a local creek. Fran was unsure of the extent to which "there were people in the community who were interested" in contributing to a learning community, and explained that she understood that "people are making different choices…partly because lives are busy…they feel they have to [make choices], and maybe they really, really do have to [so] it's hard for them to want to [for example] look after the creek". Fran admitted that despite her belief in the role of community, she also didn't always participate in community activities, and had not "always been to all the working bees at school". However, she did believe that if the school was to take the initiative and, for example, "adopt the creek", it would be possible to find people in the community "who would love" the opportunity to share in such projects.

5.7.2 Simon

Simon had been teaching for 6 years, the last 4 of which had been at West Quay Primary School as a science and ICT specialist (Prep and grades 1–6). Simon was coordinating the implementation of SSP, and had overseen the completion of the initial school energy, water and waste audits.

Understanding SSP Simon explained that SSP facilitated "schools in assessing" their operational sustainability. He described SSP as essentially "a process…a whole bunch of steps…that you go through to get accredited", and noted that the program ends when "you become a five-star accredited school". He also stated that he was not convinced that the benefits of attaining the SSP five-star accreditation were worth the time and effort required to implement the program.

Despite coordinating the implementation of SSP, Simon did not wish to be thought of as an "environmentalist", stating that "I'm not really slanted in that direction". He explained that, for example, he would certainly not "rally if they don't shut the Gordon River Dam" and "I'm not going to freak out because they're cutting trees down". He was also concerned that the program might contradict his belief that "the needs of people" must be "weighed very carefully" against the needs of the environment, but then added "I'm interested in that [sustainability] now, from a really balanced kind of angle".

Simon thought that the ultimate aim of SSP was "attitude and behaviour change". He described the role of teachers as trying to "instil those [sustainability] attitudes in kids" through "all the teaching and behaviours that you see in the school" as a result of implementing the program. Despite describing this aim as "really important [and] a good cause actually", Simon was concerned that exposing students to the enormity of environmental issues could "make them really fearful", particularly in the absence of positive ways in which they could act in response to such concerns. He therefore considered it most important to keep the content and context of SSP relevant to the age and ability of students.

SSP in Practice Simon chose a Grade 4–5 ICT lesson to demonstrate the manner in which he incorporated SSP ideals into his teaching. This was one lesson from a unit in which his students were to create a 'mini' movie to demonstrate their learning about a sustainability issue discussed in previous lessons related to waste, water and energy. Simon explained that during previous lessons he had guided the students through the process of developing storyboards and scripts. The students were now ready to be instructed on how to create their movies.

As the students entered the room, Simon insisted that they take a seat in front of a computer and remain silent. Once all of the students were seated, Simon explained that during this lesson they would learn the correct way in which to create their movie. Simon then proceeded to give step-by-step instructions for the use of the movie-making program installed on the school computers. The students were expected to listen in silence and act in unison as specifically directed by Simon. The students were denied any opportunity to explore the software or engage in peer learning. All of the students' questions were to be directed to Simon. The students seemed disengaged and regularly disobeyed instructions in ways that demonstrated a confidence with the software far beyond the level of instruction that Simon offered. Simon's pedagogical approach to this lesson is represented in Fig. 5.6.

Simon noted that he had carefully facilitated discussions about waste, water and energy in previous lessons in order to maintain a positive approach so that his students could embrace sustainable actions without becoming anxious about the future. For example, when discussing water conservation, he ensured that his students were given simple and realistic ways in which to contribute, such as "brushing their teeth and turning the water tap off". He also framed environmental issues in terms of cycles, so that, for example, the students would view the current drought as a natural climatic cycle that would eventually end. Similarly, Simon stated that he explained to his students that "even if it (climate change) is just a cycle, it's still a good idea to say, build a house with insulated walls, to not buy a big [type] of a car" in the hope that his students "can leave school and take those thoughts with them so they can make good decisions".

Understanding Pedagogy Simon held strong views about the most appropriate and effective pedagogy in a primary school classroom. He did not believe that primary school students were capable of directing, negotiating, or even contributing to their learning in meaningful ways, and stated that:

> I tend to like more formal education...I find that if you give kids too much fluff they don't get a lot done...I think you've got to lead kids...have a goal...if you run it properly...I think you can get kids to arrive at your goals.

He believed that a teacher's role was to "instill" appropriate knowledge, attitudes and behaviour by directing both the content and the learning process. He strongly supported outcome-oriented education with specific and measurable curriculum goals against which the validity of assessment activities could be judged. He described his preferred pedagogy as one in which the teacher provides "everything they [the students] need" in a "framework" through which all of the students are guaranteed to "be getting definite outcomes in the curriculum".

Simon explained that the most important aspect of this style of teaching was the necessity "to come into school and know exactly what you're doing". He appreciated the rigorous planning required for such a pedagogical approach, noting that "the more planning that goes into something the easier it looks...it's nice to be prepared". He explained that the socially-critical approach of SSP did not allow a teacher to appropriately plan for student learning:

> I think if you had heavily planned this [socially-critical scenario] and have all the people lined up so that the outcome was fairly certain...or the range of things that could be done was fairly certain, well it would be fine, but if teachers just go out and wing it, the kids just I don't think are capable.

Simon indicated that because SSP aimed to facilitate behavioural change by increasing student awareness of environmental topics, he had worked to incorporate such topics into existing learning units. For example, he incorporated the concept of solar energy in a unit on the sun. This involved "giving the kids motors and solar panels...letting them wire them up and take them outside and watch the motors run". He noted that such content changes definitely influenced his pedagogy, because "I certainly wouldn't have taken the kids on a walk to show them a house with solar panels on it if we weren't doing sustainability". He stated that he liked the idea of his students undertaking some learning "using the outdoors...out into the thing [environment] they're learning about" and "taking advantage of experts in their field".

Impediments to Socially-Critical Pedagogy Simon found the open-endedness of the intended learning outcomes for SSP "a bit overwhelming", stating that "I find it a little bit difficult when I can't put things in a box, and that might be who I am as a teacher". He suggested that he could most improve his teaching of SSP if he had "a pool of resources to draw from, whether they're on a shelf or whether they're just ideas...ultra organised cupboards with lots of stuff in them", or perhaps "a web site where there are teaching ideas" specifically directed towards SSP. He acknowledged the irony of his desire to have access to teaching ideas when he had found implementing SSP to be most difficult because "it was a bit like the internet, there were probably too many ideas...too many directions". He went on to explain that the most important aspect of the professional assistance from the CERES SSP facilitators had been advice about "where we [the school teachers] could start".

Simon's vision for the continuing advancement of SSP at West Quay Primary School was the development of a shared "community and school garden". He believed that additional funds could assist with this as "if you could just snap your fingers and have a 40,000 litre water tank...that'd be nice".

5.8 The State of Play

The stories told in this chapter reveal important aspects of the "context-dependent knowledge and experience" of educators who were required to implement SSP ideals through a socially-critical pedagogy (Flyvbjerg 2004, p. 421). Table 5.1 identifies the educators who contributed to these stories, and the status of the implementation of SSP in the schools in which they were working. Each story represents a unique perspective of the rhetoric of SSP, the characteristics of a socially-critical pedagogy and the reality of incorporating these into effective classroom practices. The observations of each teacher's lesson provided a basis for data analysis, discussed in Chap. 6, to define both the rhetoric and the reality used to define an educational rhetoric–reality gap in the context of the implementation of a socially-critical pedagogy as part of SSP.

Table 5.1 The state of play: schools and individuals implementing SSP

School[a]	SSP status	Teacher[a]	Responsibilities	Observed pedagogy[b]	Support[c]
East Valley Primary School	First year	Anita	Generalist teacher, Prep	Vocational/neo-classical	Generally unsupportive
		Robyn	Generalist teacher, Prep	Vocational/neo-classical	Generally unsupportive
		Julia	Generalist teacher, Grade 1–2	Liberal-progressive	Generally unsupportive
		Karen	ICT, Prep and grades 1–6 SSP, Prep and grades 1–6	Socially-critical	Generally unsupportive
Mountain Primary School	Five-star accredited	Elizabeth	Generalist teacher, Grade 3 SSP coordinator	Vocational/neo-classical	Generally unsupportive
		Stephanie	Kitchen garden manager		
Ocean Primary School	Five-star accredited	David	Generalist teacher, Grade 3 Curriculum manager, grades 3 and 4 Teacher mentor SSP coordinator	Liberal-progressive	Highly supportive
Sirius College	Five-star accredited	Cathy	Generalist teacher, Grade 3 Middle school coordinator SSP coordinator	Socially-critical	Highly supportive
South Bay Primary School	First year	Helen	Principal		
		Lisa	Generalist teacher, Grade 5–6 Science, grades 3–6	Vocational/neo-classical	Supportive
West Quay Primary School	First year	Fran	Generalist teacher, Grade 4–5 LOTE, Prep and grades 1–6 OHS, whole school	Vocational/neo-classical	Supportive
		Philip	Principal		
		Simon	ICT, Prep and grades 1–6 Science, Prep and grades 1–6 SSP coordinator	Vocational/neo-classical	Supportive

[a]Pseudonyms are used for all schools and all teachers
[b]The pedagogical style of observed classes is discussed and defined in Chap. 6
[c]A teacher's subjective perception of the level of support for their efforts to implement SSP

References

Allen, M. J. (2004). *Assessing academic programs in higher education*. Bolton: Anker.

Flyvbjerg, B. (2004). Five misunderstandings about case-study research. In C. Seale, G. Gobo, J. F. Gubrium, & D. Silverman (Eds.), *Qualitative research practice* (pp. 420–434). Thousand Oaks: Sage.

Huba, M. E., & Freed, J. E. (2000). *Learner-centered assessment on college campuses*. Needham: Allyn & Bacon.

Kemmis, S., Cole, P., & Suggett, D. (1983). *Orientations to curriculum and transition: towards the socially-critical school*. Melbourne: Victorian Institute of Secondary Education.

Chapter 6
The Rhetoric and the Reality

In order to find ways in which to reduce the development of educational rhetoric–reality gaps when Education *for* Sustainable Development programs, such as the Sustainable Schools Program, are introduced into schools, it is essential to first understand the rhetoric and the reality that actually defines a rhetoric–reality gap in an educational context. This chapter explores both the rhetoric and the reality of teachers' classroom practices as they responded to the requirement to implement the socially-critical pedagogy of the Sustainable Schools Program. This includes the rhetoric used by each teacher to explain their understanding of the educational and environmental goals of the program, and the reality of the manner in which they attempted to achieve the goals they identified.

This chapter also highlights the importance of an ontological-in-situ framework, informed by Giddens' theory of structuration, in identifying the critical elements of the duality of structure and agency which underpinned the relationship between each teacher's rhetoric and the reality of their classroom practices, including: permission and support; knowledge required to implement the Sustainable Schools Program; the need to implement a socially-critical pedagogy; and previous teaching experience. Both the rhetoric and the reality used to define a rhetoric–reality gap in the context of the requirement to implement a socially-critical pedagogy as part of the Sustainable Schools Program is identified.

6.1 Rhetoric

Rhetoric can provide valuable insights into how the ontological elements of structure and agency contribute to the development of educational rhetoric–reality gaps. In terms of implementing the Sustainable Schools Program (SSP) and the associated socially-critical pedagogy (see Figs. 4.1 and 4.2), the rhetoric used by the principals to justify their decision to implement the program in their schools and their

© Springer International Publishing Switzerland 2016
J. Edwards, *Socially-critical Environmental Education in Primary Classrooms*,
International Explorations in Outdoor and Environmental Education 1,
DOI 10.1007/978-3-319-02147-8_6

expectations of it, and the rhetoric used by the teachers to describe the educational aims of the program and to justify their classroom practices, is most important. The analysis of the principals' and teachers' rhetoric revealed the similarities and differences between the educational aims and the pedagogical guidelines specified by SSP, and the manner in which the principals and the teachers understood and interpreted those aims and guidelines. This highlighted the manner in which Giddens' 'structured sets' may have influenced the development of educational rhetoric–reality gaps in the teachers' implementation of SSP.

6.2 The Rhetoric of Principals

Giddens' notion of the duality of structure and agency, represented in the ontology in-situ framework (Figs. 4.1 and 4.2), suggested that teachers' practices both influence, and are influenced by, the structural elements of their educational work environment. In addition to the rules and policies outlined in SSP documents, major structural elements of the work environments of the teachers arose from the role of principals as responsible for the decision to implement SSP. Understanding the principals' motivations was therefore critical for understanding the manner in which they influenced SSP implementation in their schools, and the effect of their expectations on the teachers' practices in terms of the development of rhetoric–reality gaps. The analysis of the rhetoric of principals provided valuable insights into several of Giddens' (1979, 1984) structural elements, most particularly: hierarchical management systems (structure); school educational aims (structured principles); established school processes (structural properties); and ideologies of different school groups (social systems of interaction; see Fig. 3.4). These structural elements are discussed in relation to two main themes that emerged from this analysis; the principals' understanding of the 'aims of SSP and the purposes of education' (Sect. 6.2.1), and 'SSP as a vehicle for change' (Sect. 6.2.2). The manner in which the 'structured sets' of SSP, structural elements, and principals' motivations and actions interacted and influenced the implementation of SSP is outlined in terms of 'principals and SSP' (Sect. 6.2.3).

It is important to note here that only two principals, Philip (West Quay Primary School) and Helen (South Bay Primary School), agreed to speak on record about the implementation of SSP in their schools. The principal of Mountain Primary School had no knowledge of SSP, stating that the previous principal had made the decision to implement it. The principal of Sirius College was not available, but indicated that any of the teachers, including Cathy, could accurately explain why the program had been implemented at the school. The principals of East Valley Primary School and Ocean Primary School both explained that, because the implementation of SSP had created significant division amongst the teachers, they did not feel that they should formally discuss aspects of the program. However, both principals agreed that the teachers were free to choose to talk about their efforts to implement the program. Philip and Helen offered a range of ideas regarding their interpretation of the role of

SSP in directing both learning and teaching within their schools. Informal conversations with principals from other schools, in addition to reports from teachers and anecdotal evidence from the staff at the Centre for Education and Research in Environmental Strategies (CERES), indicated that the ideas presented by Philip and Helen were broadly representative of other principals' perspectives of SSP.

6.2.1 Aims of SSP and the Purposes of Education

Stevensen (1987, 2007) argued that rhetoric–reality gaps in the practices of environmental education "should be expected given the traditional purposes and structures of schooling in western industrialized societies" (2007, p. 129). In light of this, it was important to consider the relationship between the decision to implement SSP in schools, and the principals' and teachers' understanding of the purposes of education.

The principals of the schools were responsible for the decision to implement SSP. According to structuration, such decisions represented one aspect of the hierarchical management system, a structural element, of schools which directly impacted on the work environments of the teachers. Principals indicated that their decision to implement SSP stemmed mainly from a desire to develop more-effective environmental education. They believed that such education should be future-orientated and focused towards influencing the predominant human–environment relationships of society. Helen, for example, described SSP as education which aimed to ensure that the natural environment, and the human life that it supports, will still be "here in 100 years". Philip compared SSP goals to social responsibility. He referred to the responsibility of educational institutions to identify ways in which to actively address social and environmental concerns, in addition to their responsibility to "produce responsible, ethical individuals". He stated that "I don't think people are as aware of the environment and their impact on the environment as they could be", and that "people just can't afford to ignore these kinds of things any more". He viewed SSP as a "just cause" that "needs to be done", and from which society, not just his school, would reap "great benefits". These comments reflected the principals' practical consciousness (see Fig. 3.4). They indicated that these principals' believed that not only did society need to address certain environmental concerns, but that schools also played a vital role in ensuring that this occurred. These principals' ideas of 'social responsibility' with respect to the effect of human behaviour on natural and social environments correlated well with the stated goals of SSP.

Teachers from other schools reported that similar ideas contributed to SSP being implemented, but that these were often justified in terms of the school's responsibility to be 'seen' to be caring for their 'own' natural environments (school grounds). In other words, their principals' discursive consciousness, or verbal justification (e.g. Fig. 3.4) related the decision to implement SSP to the needs of the school rather than the needs of society or the environment. Both Cathy and David indicated that the natural environments of their school grounds represented a physical resource which strongly influenced the decision to implement SSP. Cathy for example, stated that "the management team here at the campus have always been interested in developing a program that was environmentally focussed...because of our [school] environment". Similarly, David reported that his "school [had] a very large site...so we felt we had to be environmentally responsible about what we did with it". These comments indicated that not only did the principals use their schools' natural environments to justify implementing SSP, but that, in line with Giddens' notion of the duality of structure and agency, the schools' structural elements, in the form of the natural environmental or physical resources, also influenced the principals' decisions, or agency.

The rhetoric of the principals, and the teachers, also provided valuable insights into the purposes of education which underpinned the principals' decisions to implement SSP. A comprehensive review of the purposes of education, a subject of continuing and vigorous debate, is beyond the scope of this discussion. However, Labaree (1997) provided a useful framework for analysing the principals' rhetoric and actions. Labaree grouped the purposes of education into three main outcomes: "democratic equality" which prepares students to embrace the "full responsibilities of citizenship"; "social efficiency" which prepares students for their "economic roles"; and "social mobility" which provides students with a "competitive advantage in the struggle for desirable social positions" (Labaree 1997, p. 42). The school principals described the outcomes of SSP in terms of "social responsibility" (Philip), a notion that correlates well with Labaree's (1997) education *for* democratic equality.

The manner in which the principals linked SSP to specific purposes of education was not entirely unexpected, and demonstrated a good understanding of SSP goals. As outlined by the Ahmedabad Declaration (UNESCO 2007), the social transformation needed to counteract the detrimental effects of current human–environment relationships requires urgent changes to the purpose and practices of education. This idea was also supported by the findings of a 2009 survey of the principals of Australian primary schools who were asked to identify what they considered to be the most important "purposes of schooling" (Cranston et al. 2009, p. 2). Here, the need to not only "help students develop a love for learning" but importantly to also "develop [the students'] capacities to become active and responsible members of democratic society" were identified as the two most important purposes (Cranston et al. 2009, p. 3). Thus, the reasons identified by the principals for implementing SSP reflected an understanding of the purposes of education, including the notion of education *for* democratic equality, shared by primary school principals across Australia.

However, the notion of 'democratic equality' as a purpose of education is somewhat problematic, despite being recognised by many researchers as an essential component of educational goals (Carneiro et al. 2006; Feinstein 2000; Margo et al. 2006). A review of Australian educational policies showed that since the 1980s, policies have focused on purposes most representative of Labaree's (1997) notions of social efficiency and social mobility, almost to the exclusion of democratic or public purposes (Cranston et al. 2010; Mulford and Edmunds 2010). Cranston et al (2010) reported that although recent policy documents, such as the Melbourne Declaration on Educational Goals for Young Australians, identified the need to develop "active and informed citizens" (MCEETYA 2008, p. 9) this purpose was lost amongst a plethora of statements that prioritised social efficiency and social mobility purposes. In addition, despite identifying some rhetoric congruous with democratic purposes in educational policies, the reality of these purposes being implemented had not yet been achieved (Cranston et al. 2010). Mulford and Edmunds (2010) agreed, stating that:

> the large number of expectations on schools and especially the current emphasis on the private purposes of education is unhealthy for Australian society, not least because it runs the danger of producing self interested, competitive and culturally bound individuals who are more interested in their own self advancement than they are in making a contribution to the common good. In a globalising world where the role of the nation-state is changing and societies are becoming increasingly culturally diverse, schools are needed more than ever for the important public purpose of forming active citizens for democratic publics - people with the will and commitment to shape, and participate in, an inclusive and democratic civil society and polity that are responsive to the new environment (p. 2).

The principals positioned democratic equality as an important purpose of education by choosing to implement SSP in their schools. These principals identified the ability of SSP to facilitate educational goals not strongly advocated by government curriculum guidelines, and therefore potentially not the main focus of the well-established classroom practices of their teachers. This not only confirmed these principals' practical consciousness' regarding 'social responsibility' and their belief in their social role in the provision of education, but also highlighted that these principals wanted change, and considered the implementation of change to be an important part of their role as a principal.

The changing role of school principals is the focus of a growing area of educational research, particularly in relation to leadership and management (Robinson 2006, 2007; Sahid 2004). Much of this research has identified the need for schools to re-define and re-design themselves in order to effectively meet the challenges of a world in which change is the norm (Beare 2001). The role of a school principal is seen as vital to the success of these changes, and in the context of this research, particularly in terms of providing leadership that will improve student learning. Although several styles of school leadership have been identified and described (Robinson 2006; Watson 2009), "the more leaders focus their professional relationships, their work and their learning on the core business of teaching and learning, the greater their influence on student outcomes"(Robinson 2007, p. 12). Both Helen and Philip acknowledged the need for whole-school change, but importantly, also the need for pedagogical change. These principals aimed to influence the learning outcomes for students in the schools through pedagogical change.

6.2.2 SSP as a Vehicle for Change

The principals viewed SSP as an appropriate framework for developing the future-oriented and social transformative education they desired, because it encouraged the development of opportunities for authentic learning. Helen noted that SSP provided a "good life experience for children in terms of connecting with the real world". She believed that incorporating SSP ideals within the curriculum provided opportunities for students to engage with "real life learning" by linking "real life to essential learning", particularly in terms of establishing realistic contexts for developing basic skills in literacy and numeracy. However, both Helen and Philip understood that achieving such ideals required significant school-wide change. For these principals, SSP was most valued as "a vehicle for whole-school change" (Philip). This suggested that these principals believed that mandating a set of rules and policies (a structured set) would lead to change by influencing the teachers' practices, or in other words, that structure would strongly influence the teachers' agency.

The need for change was of significant concern for the principals. Although Philip referred to the need to make people "aware of the environment", he did not equate the SSP with traditional fact-based vocational/neo-classical pedagogy. Instead, he described SSP "as a curriculum development vehicle"; a way in which to alter the traditional teacher-centred, content-based teaching practices that were entrenched within all levels of the school. The teachers from other schools agreed that SSP had been implemented for its potential to foster school-wide pedagogical change. Julia for example, stated that her principal "wanted to point us into a direction, because we needed a direction…we were floundering without her sort of leadership" due to entrenched and outdated pedagogical practices. Similarly Anita stated that her participation in the implementation of SSP was because "we're doing inquiry-based learning right through the school". Anita explained that her principal was guiding the teachers in undertaking inquiry-based pedagogies, as a transition from vocational/neo-classical practices to the more socially-critical practices advocated by SSP.

Philip noted that, like many other schools beginning to implement SSP, "we have a very traditional…teaching method, and a lot of our…planning and curriculum development hasn't changed in an eon…[there's] been a big push here…to change, but it's something that's very hard to do", and that even though the recently introduced government curriculum guidelines (the Victorian Essential Learning Standards, VELS), required teachers to move away from these traditional practices, the majority of the teachers had not responded. These comments suggested that despite implementing SSP in order to facilitate pedagogical change, Philip's own experiences had shown that mandating a new set of rules or policies (structured set) alone did not guarantee that teachers would alter their practices.

Helen agreed that SSP represented an opportunity for change as it was a "perfect vehicle in terms of [complying with] VELS", by providing "teachers with a hook, a new way of teaching with a new way of learning". Helen believed that the teachers would be motivated to embrace this "new way of teaching" not simply due to the new rules and policies outlined in SSP documents, but in response to changes to the operational practicalities of the school. Helen summed up SSP as "much more than environmental education...it's not an environmental program...it's an effective school model". Helen's comments provided insights into how the principals understood that SSP differed from traditional curriculum programs in facilitating change. Implementing SSP required making significant changes to the manner in which a school operated, which in turn, altered many of the structural elements of the teachers' work environments. Helen believed that such wide-ranging changes would be more likely to influence the teachers' practices than simply the provision of a new curriculum document.

The notion that specific educational outcomes require specific pedagogical approaches was identified by Cranston et al. (2009) as an understanding held by the majority of Australian primary school principals. When surveyed, the principals suggested that the most important strategies for achieving the democratic purposes of education included encouraging "students to accept responsibility for their own actions", making "students the focus of what happens in schools" and encouraging "respect and cooperation among students" (Cranston et al. 2009, p. 5). David for example, explained that the principal and teachers of Ocean Primary School chose to implement SSP as a vehicle through which to improve students' basic life skills. They believed that a socially-critical pedagogy had the potential to assist the students to become independent citizens with the skills required to make important life decisions, outcomes not generally linked to vocational/neo-classical pedagogies (Kemmis et al. 1983). Perhaps in recognition of the ubiquitous usage of vocational/neo-classical pedagogies, and the extent of the change required to replace these with socially-critical approaches, the principals also indicated that achieving such educational goals required schools to "value and foster the professionalism of teachers" (Cranston et al. 2009, p. 5). In other words, not only did the principals acknowledge that the teachers were at the forefront of pedagogical change, but that they also required assistance in order to enact new practices. Despite these understandings, Australian principals indicated that current educational practices failed most to achieve educational purposes related to assisting students to "develop a love of learning" or to "contribute to an environmentally sustainable society" (Cranston et al. 2009, p. 8).

6.2.3   Principals and SSP

The decision to implement SSP in the schools reflected the principals' belief of the importance of the purposes of education identified by Labaree as "democratic equality" (1997, p. 42). The principals understood the intended goals of SSP. They believed that the environmental focus of SSP enabled their school to embrace a sense of social responsibility, not only by moving towards a more sustainable relationship with school environments, but also by developing students' sense of responsibility and ability to influence the effects of human–environment relationships into the future. Most significantly however, the principals indicated that SSP represented a vehicle for change, not just in terms of its ultimate social transformative goals, but in terms of the educational processes required to achieve these. In general, the decision to implement SSP was directed towards providing an environment which challenged teachers to modify well-established, but outdated, pedagogies.

The decision to implement SSP indicated that the principals of the six schools discussed here, like many Australian primary school principals, acknowledged the need to develop teaching strategies to assist their students to become actively involved in society, particularly in ways that would enable them to "contribute to an environmentally sustainable society" (Cranston et al. 2009, p. 8). The decision to implementation SSP reflected the principals' acknowledgement of the presence of a rhetoric–reality gap between 'democratic equality' purposes of education in recent policy documents, and the 'social efficiency' and 'social mobility' purposes most strongly supported by well-established pedagogies. This in turn, reflected the practical consciousness of the principals who not only agreed with the future-oriented and socially transformative goals of SSP, but also believed that schools played a vital role in achieving these goals. The discursive consciousness of the principals however, was often focused more on the need for their schools to be perceived as acting according to parental and societal expectations than for the ultimate needs of the environment. The principals proudly showcased the resources developed for the implementation of SSP, particularly the money-saving energy efficiency systems and the newly developed modern outdoor learning spaces and garden areas for students. The principals of the schools in the early stages of SSP implementation expressed their desire to achieve their SSP five-star accreditation in record time.

Most significantly, the principals viewed SSP as a vehicle through which to initiate pedagogical change within their schools. This change was to be undertaken by the teachers.

6.3   The Rhetoric of Teachers

Giddens' theory of structuration, represented in the ontology in-situ framework (Figs. 4.1 and 4.2), indicates that when implementing SSP each teacher's practices would have influenced, and been influenced by: their unique but complex and dynamically interrelated values, attitudes and beliefs; their interpretation of SSP; and their perception of both the constraining and enabling characteristics of various structural elements of their work environment. The following discussion presents insights from the analysis of the rhetoric of the teachers regarding the ontological elements they identified as significant to their implementation of SSP, including their: social expectations (unconscious motives); beliefs about student-teacher relationships (practical consciousness); and their perceptions of structural elements such as the rules and policies of SSP (structured sets). These significant aspects are discussed according to three main themes that emerged from this analysis and which represented the teachers' understanding of: the 'aims of SSP and the purposes of education' (Sect. 6.3.1); 'achieving the goals of SSP' (Sect. 6.3.2); and 'understanding pedagogy' (Sect. 6.3.3). The manner in which the structured sets of SSP and the ontological elements related to the teachers' unconscious motives and practical consciousness interacted and influenced the teachers' approach to implementing SSP is outlined in terms of 'teachers and SSP' (Sect. 6.3.4).

6.3.1 Aims of SSP and the Purposes of Education

The teachers surmised that SSP had probably been developed in response to society's growing awareness of, and concern for, certain aspects of current human–environment relationships, which, according to Cathy had developed into "issues which are so important to society at the moment". In particular, these issues reflected the "way people are heading", particularly with regard to the use of "natural resources" (Cathy). The teachers related these issues to the notion that "what we've got on Earth is really limited" (Anita) and that therefore society must "take into account the fact that things [natural resources] are limited…they're finite" (Anita). In light of this, Fran described the ultimate aim of SSP as encouraging people to "try to change the way we are currently using resources" in order to "sustain the resources we currently have" and ensure that "life in the world becomes sustainable" (Lisa). The teachers believed that this required society to learn ways in which to not only "maintain the environment" (David) but also to "care for the environment" (Cathy) and to "have respect for their environment" (Karen). Such comments indicated that the teachers agreed with the principals that SSP was essentially future-oriented and socially transformative education which aimed to influence human–environment relationships, and that SSP represented environmental education based on the "environmental preservation and restoration" and "natural resource conservation" (UNESCO 2005, p. 28) components of education *for* sustainable development (ESD). In addition, this correlated well with the results of surveys of teachers undertaking environmental education (e.g. Grace and Sharp 2000; Tomlins and Froud 1994) which found that teachers associated environmental education with educational goals to develop students' "personal responsibility for the environment" and "future attitudes to the environment" (Cotton 2006, p. 69). The teachers identified their role as educators in addressing such social concerns, through the aims of SSP, by "teaching children how to save the environment" (Robyn) or assisting "this generation to be making the changes now into learning a different way of living" (Fran). This role was seen to incorporate helping children to learn different ways in which to "use our environment" (Cathy) and to initiate "changes that children can make now that will have long term positive effects" (Fran). Julia for example, described her role as teaching ways of "not being wasteful". The teachers agreed that SSP was about "reinforcing the need for us to change the way we are currently living on the planet" (Fran), and therefore, "in the long term make changes to the world" (Karen). The teachers agreed that the ultimate goal of SSP was to influence human behaviour to ensure a more sustainable use of the Earth's natural resources.

These comments provided valuable insights into the teachers' beliefs about the aims of SSP, the purposes of education, and their roles and responsibilities as SSP educators. The teachers understood that SSP was not part of the current standard curriculum, but that it had been designed and introduced in response to public concerns. This suggested that the teachers' recognised SSP as education that had been "socially constructed" and which supported the notion that "purposes for the next decade can only be based in our current circumstances and our preferred futures" (Schofield 1999, p. 9). Not only did this correlate with the principals' understanding of the goals of SSP, but also their beliefs regarding the purposes of education. The teachers agreed that SSP was a program "concerning the good of society" (Schofield 1999, p. 14) and which therefore represented education *for* democratic equality (Labaree 1997). These ideas reflected the teachers' practical consciousness (see Figs. 4.1 and 4.2), and indicated that these teachers, like their principals, believed that not only did society need to address certain environmental concerns, but that schools also played a vital role in ensuring that this occurred. The teachers' ideas regarding the need to reduce society's wasteful overuse of natural resources correlated well with the stated goals of SSP, and framed SSP in terms that were both suitable and accessible to the students of primary schools.

6.3.2 Achieving the Goals of SSP

The teachers understood that the goals of SSP would be best achieved through the use of specific instructional practices which provided certain types of student learning experiences. Although many of the teachers referred to the need for students to "learn about the environment" (Julia), indicative of an education based on knowledge acquisition, it was clear that this was not what most of the teachers intended. David for example, pointed out that SSP is "not a subject" and that "the environment isn't a subject". The teachers envisaged SSP goals as being achieved through an educational process which provided opportunities for students to develop their awareness of current human–environment relationships. This included, for example, ensuring that students understood the "concept of what waste is" (Julia), by "trying to get children to understand how to use less water, and less energy" (Robyn) through exploring the notion of "reduce, re-use and recycle" (Julia). Although several of the teachers indicated that "being more aware" (Elizabeth) represented a greater knowledge *about* the environment, most related this awareness to understanding the ways in which individuals and societies are interrelated. Anita for example, indicated that her students should develop an awareness of the ways in which the "choices we [individuals] make can impact on other people's choices" (Anita). Several of the teachers linked the notion of increased awareness to the acceptance of responsibility. Fran, for example, described SSP learning activities as helping her students to become aware of the notion "that we [humans] have a responsibility to change the way we're currently operating for environmental reasons". The teachers felt that the ability to "feel some sense of responsibility" (Lisa) was necessary for their students to eventually "take ownership of how they affect the environment" (Andrew). Cathy noted that student awareness, in this sense, was "not only an idea of basic knowledge about the world, but [also] how they [the students] can make a difference". As such, the teachers suggested that the SSP learning activities aimed to facilitate their students' understanding "that taking action is so important and [that] just little bits [actions] make such a big difference" (Cathy). David referred to broader goals related to "our social responsibility and our civics and citizenship work" which assisted his students to become active citizens who were "not just taking up the environment and wasting space or using up the air, but actually contributing [and] actually making a difference". David linked this to the notion that societal changes supported by SSP were best achieved through "building better pillars of society…[students who would leave] the community in a better way than [they] found it". In other words, the teachers indicated that developing their students' awareness and sense of responsibility would facilitate behavioural change. This corresponded with the results of other research in which surveys of teachers undertaking environmental education (e.g. Grace and Sharp 2000; Tomlins and Froud 1994) found that teachers associated environmental education with educational goals to develop students' "personal responsibility for the environment" and "future attitudes to the environment" (Cotton 2006, p. 69).

These comments provided valuable insights into the values, attitudes and beliefs, which, in line with the ideals of Giddens' theory of structuration, and as incorporated into the ontology-in-situ framework (Figs. 4.1 and 4.2), interacted to form the unconscious motives of many of the teachers. There was a general consensus that implementing SSP was a matter of 'social responsibility', not only in terms of the responsibility of the teachers implementing a curriculum that was derived from social concerns, but also in terms of the role of the teachers in developing their students' understanding of their social responsibility. It seemed that the teachers accepted the notion that society expected them to undertake such a role, particularly as SSP had been developed in response to social concerns. In the same way that the principals identified the need to be perceived to be addressing social expectations regarding, for example, the care of school grounds, the teachers indicated that they felt the need to undertake their role in a manner that met perceived social expectations, that is, social norms.

Most of the teachers referred to the importance of increasing their students' 'awareness' of human behaviour as the first step in developing a sense of social responsibility in relation to the human–environment relationships addressed through SSP. They believed that this would encourage students to fulfil their social responsibility by altering their own behaviour. In other words, the teachers correlated looking after the natural environment by teaching for behavioural change, with social responsibility. However, some of the teachers suggested that behavioural change would occur only if it was accompanied by a change in attitudes. They referred to the need to "try to instill those attitudes in kids" (Simon), but did not agree on which specific attitudes they should assist their students to develop. Cathy, for example, aimed to develop respect, stating that "I'm trying to teach the word respect…respect for themselves, respect for the environment", whereas Lisa concentrated on attitudes which contributed to "all that consumerism". Lisa believed that SSP goals required her students to learn that they:

> don't need everything brand new and that sort of attitude [be]cause kids [think] everything's [got to] be new, and everything's [got to] be up to date so they've [got to] have every different 'Game Boy' play station or whatever it is going around...they can't just have one, they've [got to] have more.

Fran described the need to instill these new attitudes as habits, referring to the need to "change habits and instill new habits", in order to "make these responses [new attitudes] automatic" so that they "become the norm".

The teachers believed that, due to the focus on changing students' attitudes and habits, SSP aimed to position students as agents of behavioural change beyond the school grounds. Elizabeth noted, for example, that although SSP "starts in the classroom with education" the students take their new understandings "back to the family so that it's just not a 9 to 4 concept...it's a 24 hour concept...it's not just a school curriculum...it's an outreaching...program as well". Lisa suggested that SSP engaged her students in ways that encouraged them to develop a "passion to go away and learn something that they need to be putting in place at home and around the community". In this manner, students were actively "imparting [their new understandings] to someone else" (Elizabeth) and could, for example, "help change the habits of [their] parents" (Robyn). Similarly, Cathy related attitudinal change to the ability of students to maintain their roles as change agents as they developed their roles in society, stating that "I'm hoping if they start to change their attitudes, it will have this fantastic flow-on effect as they start to get older". In other words, the teachers viewed SSP as working towards social transformation through changing the attitudes and behaviour of their students.

These comments reflected the teachers' understandings of two important aspects of human behaviour and their relationship to social norms: the values and attitudes that direct an individual to behave in certain ways; and the habits, or behavioural routines that position certain behaviours as social norms. Recognition of the importance of an individual's values and attitudes to their actions and understanding of value-laden environmental issues is paralleled by studies indicating a pervasive belief amongst primary school teachers that environmental education must include the teaching of attitudes (Cutter and Smith 2001b). Similarly, an Australian Government study commissioned in 2002 identified the values considered by school communities to be the most critical components of values education in schools, including: tolerance, respect, responsibility, social justice, excellence, care, honesty, freedom, and being ethical. The maintenance and preservation of the natural environment was specifically mentioned as a part of each individual's responsibility (DEST 2005).

However, this type of education is somewhat problematic, and would require teachers to determine what values and attitudes are, how they are constructed, and which values and attitudes should be taught in order to achieve a certain aim. In addition, although the notion that behaviour is strongly linked to an individual's values and attitudes is strongly supported in the literature (e.g. Feather 1992;

Gynnild 2002; Raulo 2000; Rohan 2000), research indicates that the ability of one individual to 'instill' a new attitude on another is, at best, difficult, if not impossible (Doll and Ajzen 1992; Fishbein and Ajzen 1975). Irrespective of whether or not any teacher could facilitate value or attitude change in their students, the perception of the strong relationship between values and environmental issues has led to concerns by some educators that incorporating environmental ideals in any curriculum is simply indoctrination (e.g. Burbules and Berk 1999; Jickling and Spork 1998). Certainly some aspects of the rhetoric of the teachers, particularly that of Simon and Fran, who described the need to 'instill' attitudes and habits in their students, might be interpreted as indoctrination. However, a closer look at the rhetoric of most of the teachers did not indicate that they viewed SSP or socially-critical pedagogy as indoctrination. They referred to the need to develop their students' sense of "social responsibility" (David) and "passion to go away and learn" (Lisa). Specific values and attitudes mentioned by the teachers included the need to learn "respect" (Cathy), to take "ownership" of their actions (Andrew), and to be "less wasteful consumers" (Julia), in other words, to become more-aware citizens. These attitudes and values were representative of those considered by school communities to be critical components of school education, discussed above (DEST 2005), and reflected the notion that the goals of SSP were strongly related to the purposes of education identified by both the principals and the teachers to be associated with democratic equality. In addition, and as argued by many educators (e.g. Fien 1993; Huckle 1986; Kelly 1986; Scott and Gough 2003) it is the intentions and manner in which teachers approach their teaching, not simply the style of pedagogy, that determines when education becomes indoctrination.

Although a teacher's ability to 'instill' specific values or attitudes in their students is uncertain, their ability to encourage their students to develop certain habits, or behavioural routines, is more achievable. In line with Giddens' notion of the duality of structure and agency, assisting students to behave in certain ways in response to current social norms will in turn influence those social norms. Although such a behaviourist teaching and learning focus may encourage little more than "green consumers" (Gayford 1996) the teachers believed that the development of appropriate behavioural routines could encourage their students to act as change agents in their families and local communities (of course, behavioural routines were also an important component of the teachers' practices; see Sect. 7.2).

However, the teachers noted that the ability of their students to effectively influence society in this manner required more than just attitudinal and behavioural change and that it was essential for their students to also develop a range of both practical and thinking skills. This understanding was supported by the idea that creating sustainable human–environment relationships through education is, in fact, a process to "learn how to learn and how to be critical" (Scott and Gough 2004, p. xiv). Anita, for example, suggested that her students needed to begin to think "about the kinds of things they can do at home, the kinds of choices" that were available to them. Similarly, David indicated that influencing others required "being able to talk about it [a behavioural change] intelligently" in order to explain "why you're doing it and why it's important". Fran described this as the need for her students to develop the confidence or "the guts to question".

David suggested that if students were to develop these types of skills, implementing SSP required an "integrated approach" in which both the teachers and the students engaged with a range of "themes which are a part of life". This correlated with Helen's reason for implementing SSP: its ability to interest students in learning through participating in real life. Karen on the other hand, indicated that achieving SSP goals depended on students being able to understand and consider the repercussions of a history in which humans "have not respected our environment", and that this required the teachers to assist their students to develop an "understanding of what it's been like, how we've used it [the environment], and now what we've got to do to make it last". Andrew agreed, stating that he aimed to:

teach students not just more about the environment but the effects…not just that they [the students] are having on the environment, but the human race…about how humans are affecting the planet…and how in the future that's really going to have some consequences for us [all humans].

Lisa related all of these ideas to "educating the children about the future", and that therefore, SSP aimed for "long-lasting" educational outcomes.

These comments suggested that the teachers understood the notion that educational goals depended on the educational process, or pedagogy, and that therefore SSP goals required teachers to develop specific learning experiences for their students. The teachers indicated that these experiences needed to provide opportunities for their students to, amongst other things, critically assess the historical perspective of the social concerns addressed by SSP, to critically evaluate their own role in these concerns, and to imagine a range of possible futures. In addition, appropriate learning experiences were needed to build their students' confidence and to develop their skills to question not only what they see, but also what they believe and what they imagine. Thus, analysis of the rhetoric of the teachers indicated that they agreed with the principals that a traditional vocational/neo-classical pedagogy could not effectively achieve the goals of SSP. The teaching strategies and learning outcomes identified by the majority of the teachers as the best way in which to achieve the goals of SSP most strongly correlated with a socially-critical pedagogy, through which learning experiences could provide their students with opportunities to gain "historical and critical perspectives on society" and to "engage in activities that are consistent with building a responsive democratic society" (Gough 1997, pp. 91–92). In other words, the teachers understood that SSP required "teachers [to] shift from control of knowledge to creation of processes whereby students take ownership of their learning, and take risks to understand and apply their knowledge" (Wink 2000, p. 135). The characteristics of a socially-critical pedagogy, and the manner in which this informed SSP guidelines and activities is outlined in Chap. 2.

Although the teachers' rhetoric indicated an understanding of the need for a socially-critical pedagogy, the majority of the teachers identified aspects of student learning rather than issues of teaching as the mechanism through which SSP goals would be achieved. Both Julia and Anita did report that their principal was assisting them to alter their pedagogy in order to more effectively implement SSP, but Cathy was the only teacher to indicate that achieving the educational goals of SSP was actually dependent upon the teachers undertaking new practices. She described SSP as "trying to empower teachers, or really teaching teachers to empower children".

Cathy's understanding reflected the motivation of the principals who chose to implement SSP in order to encourage the teachers to learn to teach differently.

In order to understand the development of rhetoric–reality gaps in the teachers' classroom practices, it was essential to identify the teachers' understandings of the characteristics and roles of different pedagogies.

6.3.3   Understanding Pedagogy

In order to investigate the relationship between teacher rhetoric and the reality of their practices, it was essential to explore and compare the teachers' rhetoric in relation to different pedagogical approaches. The teachers discussed aspects of three hypothetical scenarios (see Tables 4.1, 4.2 and 4.3), each of which represented a different pedagogical approach to undertaking a potential SSP related activity. These discussions highlighted the teachers' understandings of the differences of: a vocational/neo-classical pedagogy, an approach that principals considered to be well-established but outdated; a liberal-progressive pedagogy; and the socially-critical pedagogy that was advocated by SSP and which had been identified by the teachers as providing the most appropriate mechanism for effectively achieving what they considered to be the educational outcomes of SSP. The teachers identified and discussed these differences in terms of teacher–student relationships and the classroom practices of socially-critical pedagogy.

6.3.3.1   Vocational/Neo-classical Pedagogy

A vocational/neo-classical pedagogy (see Table 4.3) was described by the teachers as a "teacher controlled" practice in which the classroom teacher was "orchestrating and conducting" student learning (David). Although Elizabeth described this pedagogy as ensuring "a very logical and progressive approach", others described this pedagogy as "closed" (Julia), "restrictive" (Anita) and "prescriptive" (Cathy). Julia commented that "this is assuming that the teacher's the expert" despite the fact that "the teacher may not know, [teachers] really don't know everything". The teachers noted that the learning activities reflected a "teacher-led focus" to learning (Lisa) where even the "questions are provided to them [the students] rather than them generating their own questions" (Julia). Anita agreed, noting that such "teacher-directed learning", where there is "not a lot of student input", is akin to "giving the children a template" within which all their learning must occur.

Cathy was most concerned that such a pedagogical approach "didn't allow for the different learning styles" which in turn, suppressed students' "expression of individuality" and ignored the benefits of the "different ways that children…come up with such brilliant things to do". Karen referred to her own experience which indicated that learning derived from teacher-initiated ideas was "just simply not as powerful as coming from the children". Similarly, the teachers considered the assessment of student learning in vocational/neo-classical pedagogical practices to be strongly teacher-directed. Anita, for example, suggested that vocational/neo-classical pedagogies were "just done by the teacher to meet [a curriculum] criteria", or to "maximise the chances of successful completion [of a curriculum outcome]", and were therefore "not about the children's outcomes". David suggested that, for many teachers, this aspect of the pedagogy was "probably conscientious because it [addressed] things like VELS". Julia was somewhat conflicted over the need to teach specific curriculum outcomes, stating that "they [the students] need to know certain things…there's stuff that you do have to…teach them…anything from VELS you do have to do". After a moment of reflection she asked rather sarcastically "is that the reason that we do it?", and noted that the use of a vocational/neo-classical pedagogy suggested that there is always only one correct answer to any question; "it almost seems like you'd get into trouble if you didn't get quite the right answer".

Karen voiced similar concerns, stating that a vocational/neo-classical pedagogy is:

> just superfluous, it's ridiculous to try and teach that way…it has no meaning…it has no meaning to the teacher because they're doing it to please somebody who's written this curriculum…therefore it's not going to have any meaning to the child.

Karen noted that such an approach excludes many important opportunities for learning. For example, she believed that although the teacher is "trying to do the right thing because they've identified these web resources, I would get the kids to identify them themselves…part of this learning should be getting there…getting onto the web themselves". Similarly, David noted that "it's great they've got a parent coming to visit to explain things but it seems to not be as empowering to kids as the others [liberal-progressive and socially-critical pedagogies], or of authentic learning as the others".

Such comments indicated that these teachers understood the essential purpose of a vocational/neo-classical pedagogy to be the transmission of knowledge from a teacher to their students, and that this knowledge was not only outlined in curriculum documents, but that it also represented unquestionable objective truths. The teachers recognised that a vocational/neo-classical pedagogy represented a positivist view that "reality is independent of the observer" (McRobb et al. 2007, p. 2), and that the teachers who employed such a pedagogy were often "focused on the role of formal education in providing teaching which may or may not result in learning" (Schofield 1999, p. 7).

6.3.3.2   Liberal-Progressive Pedagogy

In contrast to the teacher-directedness of vocational/neo-classical pedagogy, the teachers described the liberal-progressive pedagogy (see Table 4.1) as a more 'structured' approach to learning: "a pretty good structured unit" (David). All of the teachers thought that the overall context and basic learning activities of the scenario were "good" (Karen). Andrew stated that "It's a good idea this one, I'll take that on board right now".

The teachers considered the liberal-progressive pedagogy to represent the middle ground between a vocational/neo-classical and socially-critical pedagogy, particularly in terms of the degree to which the teacher aims to control the focus, design and assessment of learning activities. For example, Anita stated that "I guess what's leaping out at me with [this] scenario is that there's not a lot of student input at the start" and that therefore, like the vocational/neo-classical pedagogy, "it just seems very adult directed". Cathy also viewed the focus of the unit to be "very teacher oriented". However, she believed that the teacher did have "some idea of where the children's interests were" and noted that in her experience, "sometimes that's how topics come up…sometimes you do design units around that sort of thing, to follow the kid's interests…it's a good way of doing it if the kids show an interest in it". Karen agreed, stating that "they're [the teachers] trying to get the kids to move [that is, to engage with learning] as a result of the observations" of the students' interests. Anita, however, indicated that although she agreed that student interest was important, she believed that it was more important to take student prior knowledge into consideration when planning and implementing a learning unit. She stated that "it doesn't even really look like they've [the teachers] tried to gauge the knowledge children actually have about the topic before they've started".

The teachers noted that, unlike the vocational/neo-classical approach which concentrated on specific learning outcomes from specific fields of study, the liberal-progressive approach does begin to integrate curriculum outcomes from different subjects. However, several teachers did not believe that this particular approach to integration would be successful, and commented that it did not reach the level of curriculum integration of a socially-critical pedagogy. Karen for example, stated that "integrate should be integrate" and asked "where is their maths, where is their literacy?". She stated that "I don't like this total control where you (the teacher) say right, I'm just going to integrate that into that basket and so and so…I don't believe in that at all…I believe…the word integrate means integrate". Similarly, Cathy noted that:

> it's [the liberal-progressive hypothetical scenario] specifically going to integrate just those two areas…I find that interesting because we try, well we do manage to interweave…a lot of our maths and our English reading program, I mean it fits perfectly into this sort of thing, so I would see it being broader.

The teachers also noted that, unlike the vocational/neo-classical approach, the liberal-progressive approach did not always position the teacher as the only source of knowledge in the classroom. Instead, teachers invited guests into the classroom as a way of encouraging students to learn from others, or from experts. Julia for example, commented that "we try and plan something [guest speakers] as often as possible, [if] I know I'm no expert at this [learning outcome]…we get people who are experts if we can". Similarly, the teachers noted that bringing a range of people into the classroom provided opportunities for their students to develop a wide range of understandings. This concerned Elizabeth who believed that bringing a guest speaker into the classroom would distract her students from the main learning outcomes of a program. She explained that, if the aim of the liberal-progressive scenario was to teach the students about native plants, "I think having the guest speaker from the local aboriginal group would be something I'd put later when the kids have got much more of a firm concept of what it is that they're doing...I'd do that after they've [completed the planting learning activities] because that way they'd be more focused on [just] plants".

6.3.3.3   Socially-Critical Pedagogy

In contrast to both the vocational/neo-classical and liberal-progressive pedagogies, the socially-critical pedagogy portrayed in the hypothetical scenario (see Table 4.2) was considered to represent a "more child-centred and child-directed" practice (Karen) which was described by Fran as learning with a "worthwhile purpose attached to it". The teachers noted that this pedagogy was characterised by "teachers leading students to do quite a bit of planning" (Simon) in a learning environment where "students choose the focus of their study" (Julia) "rather than [follow] a teacher-led focus" (Lisa). Several teachers commented that student decision making was essential because when there is "student involvement...they [students] have power over it [the learning focus] and if they own it they have a different attitude towards it" (Lisa). Fran believed that a socially-critical pedagogy encouraged students to be "more involved" in their learning which led to "more possible outcomes" than teacher-directed learning. Karen agreed, noting that when students negotiated with teachers to determine their learning focus, "the children have been empowered", and that this gave "them an opportunity to show their skills through their strengths" (Julia), and the opportunity to "build on...what they already understand" (Simon). Robyn suggested that not only did the socially-critical pedagogy advocated by SSP provide students with a feeling of ownership over their learning, but that when "the whole-school's involved in things like this there's [also] more of an ownership over the school".

Elizabeth stated that unlike a vocational/neo-classical pedagogy, a socially-critical pedagogy was most importantly "all about engagement" and embracing the notion that students "like doing rather than just sitting there passively". Julia supported this observation, commenting that her students "do learn more and respond better when they're doing hands-on stuff". Julia's notion of "hands-on stuff" was identified by several of the teachers as an important contributor to the ability of a socially-critical pedagogy to engage students. Lisa suggested that any hands-on learning activity engages her students because "they like it, they enjoy it, they want to be a part of it, they [want to] know what it's about". Elizabeth noted that hands-on learning activities improved student learning due to the cooperative and collaborative learning environments in which they usually occurred, stating that "hands-on [activities] and working with a partner...and working in small groups...is a really good way of drawing out and pushing out and maximising their [the student's] own understanding of some learning". Robyn agreed, but indicated that this was most effective when "children are working with different year levels rather than just being confined to their own classroom". Robyn's observation was supported by several of the teachers who recognised that, unlike standard school classes, a multi-age learning group could provide a learning environment that was "a realistic representation of the community" (Fran). Cathy agreed, noting that a multi-age learning experience "gives the kids a really good sense of community and how they can be part of it...how different members of the community can contribute". David described multi-age learning as "authentic learning in that it's giving the kids experiences to deal in the community and deal collaboratively". The teachers considered multi-age learning to be an essential characteristic of "teaching life" (David) through the development of authentic, real life contexts and activities offered by a socially-critical pedagogy.

Several of the teachers indicated that such multi-age learning experiences are much more effective when they incorporate communities "beyond the classroom" (Julia). Robyn stated that:

> I think you need to go out of the school classroom and then you can come back and bring what the children have learnt out there and go from there...it gives the kids such a new perspective and learning as well.

David described the learning from a socially-critical pedagogy as "coming from everyone...you've got all your...stakeholders, you've got your outside community, you've got your school community, you've got your staff, you've got your decision makers like your council, and you've got your student body at various levels".

The teachers commented that a socially-critical pedagogy encouraged students to actively participate in the learning process. Julia noted that students were most likely to become actively involved in learning activities that revolved around issues that involved "not just some other country or out whoop whoop [region] that they're [the students] not related to" but issues which concerned and interested the students, and, in relation to SSP, which are "not unrelated to what we need to be doing in our own world". She described, for example, the development of native gardens in the hypothetic scenario (socially-critical pedagogy; Table 4.2) as "good stuff because it's real stuff".

The teachers noted that the "whole school approach" (Robyn) advocated by SSP incorporated many of the attributes of a socially-critical pedagogy. The 'whole school approach' provided excellent opportunities for students to learn through participation in hands-on activities undertaken by multi-age, community-based groups. David believed that SSP could provide "rich learning" opportunities which involved "collaborative learning" and "negotiating" in an environment in which "everyone's involved and where different people have got different inputs". Such comments indicated that the teachers viewed a socially-critical pedagogy as providing opportunities for their students to learn about the ways in which individuals and societies interrelated by actively participating in complex and dynamic social groups. In addition, this was supported by the notion that effective socially-critical or transformative education must enable students to actually use the understandings they are developing, that is, to learn through experiences in which their developing socially constructed knowledge is applied to the social context of life (Fien 2001; Wink 2000).

The teachers understood that in order to undertake such a socially-critical pedagogy, as opposed to a vocational/neo-classical pedagogy, they needed to take on a
less prominent role in the learning process. For example, many of the teachers
referred to the benefits of inviting community 'experts' into the classroom in order
to "use their [guests'] resources" (Andrew), and to introduce students to "new ways
of getting information" (Robyn). Similarly, Julia believed that community guests
enabled students to access information and ideas not able to be provided by their
teachers, because "teachers…really don't know everything, there's lots out there we
[the teachers] can get other people to help us with". Karen believed that opportunities for her students to gain ideas and information from multiple sources and perspectives was a critical component of effective and meaningful learning: "the
children actually get the message not just from a teacher but from other people…it's
really powerful [when it] comes from more people than just the teacher". She
explained that, unlike the vocational/neo-classical hypothetical scenario in which a
parent was asked to describe the role of a scientist to the students, a socially-critical
approach would help those students to develop a meaningful understanding of the
role of a scientist by actually working alongside a scientist. She explained that she
had attempted to do this by enabling her students to assist in a real survey of aquatic
life with a biologist. She explained that the biologist "speaks to the children when
they're standing in waders in the water" and encourages the students to think critically about their role in the activity, saying "now if you're a scientist, you don't just
stick your bucket in the water and grab it [the sample], you've got to think 'now
what location do I go to?'". Karen noted that, as a result of this level of active participation, "they [the students] come back and they believe they are scientists".

The teachers also understood that, compared with a vocational/neo-classical
pedagogy, undertaking a socially-critical pedagogy required them to more effectively integrate student assessment into the learning process. The use of multi-age
learning groups and community settings enabled the students to demonstrate their
understandings and obtain ongoing feedback in the form of "assessment not just
from their teachers, but from their local community" (Julia). Cathy noted that, in a
situation in which peers are enabled to provide feedback, "I think the kids really
appreciate that sort of thing…they listen to what [is] being spoken about by their
peers in that sort of situation".

These ideas were supported by Schofield's (1999) observation that "the 'de-
institutionalisation' of education" so that "more formal education occurs outside the
classroom" (p. 7) is a current trend in education. When Ivan Illich first proposed the
"disestablishment of schooling and the creation of learning webs" in 1971, as a way
in which to address problems associated with traditional educational practices, the
idea was mostly dismissed as "radical" (p. 7). However, research indicates that since
the introduction of "internet-based interactive learning" (p. 7), there has been a
general acceptance of the notion that there are "multiple pathways to knowledge, to
understanding, to literacy, to skills in society" (p. 8). The comments by the teachers
demonstrated their understanding that a socially-critical pedagogy embraced the
notion that valuable learning is neither restricted to the classroom, nor to the teach-
ings of a single individual. Instead, valuable learning can occur at any time, in a

variety of social and natural environments, and in response to a wide variety of resources.

The teachers' comments about the hypothetical scenarios indicated that they recognised the important epistemological differences between the vocational/neo-classical pedagogy identified by the principals as an outdated practice in need of change, and the socially-critical pedagogy advocated by SSP. Irrespective of their preferred pedagogy, each teacher understood that the vocational/neo-classical and socially-critical pedagogies required distinctly different teacher-student relationships and classroom practices. They interpreted a vocational/neo-classical pedagogy as the "transfer of knowledge in an end form to the individual [student]", and related socially-critical pedagogy to the notion that "understanding is gained by the individual [student]...when actively examining and questioning the world around him/her" (Mogensen 1997, p. 433). These understandings not only correlated well with the principals' ideas regarding the pedagogical needs of SSP, exemplified by Helen's comments regarding "real life learning", but also confirmed that the teachers believed that achieving SSP goals required the provision of opportunities for students to develop their awareness of current human–environment relationships, particularly in terms of the ways in which individuals and societies interrelated. This was not seen by the teachers to be solely about acquiring knowledge. Instead, the teachers agreed with the principals that these goals were concerned with achieving 'democratic equality' (Labaree 1997) through developing students' sense of social responsibility by "teaching life" (David). The teachers' rhetoric indicated that they also agreed with the principals that such democratic equality purposes of education are not achieved through a vocational/neo-classical pedagogy. Instead, these educational purposes demand a socially-critical approach to learning.

6.3.4 Teachers and SSP

The rhetoric used by the teachers to describe the educational aims and pedagogical requirements of SSP provided valuable insights into their perception of various ontological elements of their work environments. The teachers fully understood the environmental and educational goals of SSP, as outlined in SSP documents. They identified the development and implementation of SSP as a response to social concerns for the current state of human–environment relationships, particularly with respect to the human use of natural resources. They agreed with the school principals that SSP was a future-oriented education which aimed to transform human–environment relationships through encouraging attitudinal, and therefore behavioural, change in students, and by positioning students as agents of social change. A few teachers, particularly Julia and Anita at East Valley Primary School, and Cathy at Sirius College, demonstrated a good understanding of additional goals, held by the principals, for SSP to drive pedagogical change. Both Julia and Anita acknowledged that their principal viewed the implementation of SSP as an opportunity to initiate widespread pedagogical change, most specifically, to interrupt the well-established "very teacher-driven" and "very content-based" (Anita) classroom practices. Cathy indicated that her principal had chosen to implement SSP for its potential to support the teachers' efforts to continue to develop and improve their practices as "part of the [school] vision" which encouraged both the teachers and the students to "have a go".

The teachers also demonstrated a good understanding of the differences in epistemology, teachers' motivations, educational aims and classroom practices of a vocational/neo-classical pedagogy and liberal-progressive pedagogy, and the socially-critical pedagogy advocated by SSP and their principals. They described socially-critical pedagogy as a strongly student-centred approach through which teachers facilitated learning experiences that evolved directly from their students' questions and interests. These experiences were considered to be most effective when they incorporated activities within a community of people of different ages, different backgrounds and with different ideas. The teachers believed that such opportunities could develop their students' abilities to negotiate, cooperate and collaborate in an environment representative of the society in which they should become active citizens, and that in this way, a socially-critical pedagogy addressed the democratic equality purposes of education to be achieved through SSP. In other words, the teachers demonstrated an understanding of the principles and practices of the pedagogical approach best able to achieve the educational outcomes of SSP.

6.4   Reality

In order to understand the educational rhetoric–reality gaps that developed as a result of implementing SSP, it was important to establish the reality of the practices within the schools implementing the program. This reality was represented by the pedagogies employed by the teachers implementing SSP within their usual occupational environments, that is, their classrooms. Consideration of the role of the critical ontological elements of structure and agency in the regionalised and routinised practices of these teachers provided useful insights into how such ontological elements contributed to the development of the educational rhetoric–reality gaps.

Figures 5.1, 5.2, 5.3, 5.4, 5.5 and 5.6 show that the teachers employed a range of classroom practices in order to implement SSP. These practices ranged from a strongly teacher-centred vocational/neo-classical pedagogy to a significantly student-centred socially-critical pedagogy, as required by SSP. Elizabeth, Fran, Lisa and Simon positioned themselves as the "authority" in the classroom, who "uses directive pedagogy" for "transmitting knowledge" to students (Kemmis et al. 1983, p. 12), typical of a vocational/neo-classical pedagogy. The practices of Anita, David, Julia and Robyn are better described as liberal-progressive, although each of these teachers' practices incorporated some aspects better identified as either vocational/ neo-classical and/or socially-critical. In contrast, Cathy and Karen positioned themselves as "co-participants with students in the learning process" (Wink 2000, p. 71) during which "teachers teach less often by didactic approaches....and more often by encouraging inquiry, critical reflection and action" (Gough 1997, p. 91). Cathy and Karen ensured that learning activities were "negotiated with the students, other staff and the wider school community" in order to position students "as the agents for producing working knowledge through interaction through others in socially significant tasks" (Gough 1997, p. 91). Thus, unlike the other teachers, Cathy and Karen demonstrated an "openness to the unplanned directions that learners will take" when engaged with learning through a socially-critical pedagogy (Vare and Scott 2007, p. 198).

6.5   SSP and Educational Rhetoric–Reality Gaps

Figures 5.1, 5.2, 5.3, 5.4, 5.5 and 5.6 highlight that, in the context of the implementation of SSP through a socially-critical pedagogy, the practices of all but two of the teachers, Cathy and Karen, constituted educational rhetoric–reality gaps. The comparison of the ideas revealed through the teachers' rhetoric with the teachers' classroom practices not only identified the presence of these rhetoric–reality gaps, but also provided valuable insights into several ontological elements, related to both structure and agency, that may have constrained and/or enabled such gaps to develop. These are discussed according to four main themes: permission and support (Sect. 6.5.1); knowledge required to implement SSP (Sect. 6.5.2); implementing a socially-critical pedagogy (Sect. 6.5.3); and teacher experience (Sect. 6.5.4).

The following discussion draws heavily on comparisons between the reality of the classroom practices of Cathy and Karen, whose practices best represent the socially-critical pedagogy required to most effectively implement SSP, and of Lisa and Elizabeth, whose practices most closely represent the vocational/neo-classical pedagogy identified by the principals as outdated and most in need of reform.

6.5.1   Permission and Support

Educational change, such as the pedagogical change advocated by the principals who chose to implement SSP, can cause a great deal of anxiety (e.g. R. Evans 1996; Fullan 2001). Several researchers have shown that teachers' anxiety about change can, in many cases, be minimised when they perceive that they have the support of their principals (Carson 2007; O'Connell et al. 2001). It is therefore not unexpected that educational rhetoric–reality gaps related to the teachers' practices of environmental education might reflect "issues of whether or not the teachers feel they have permission to carry out the activities…[that] they feel constitute environmental education" (Robertson and Krugly-Smolska 1997, p. 311). The rhetoric of the teachers suggested that they sought, and valued, permission to implement SSP and a socially-critical pedagogy from their school principals, their work colleagues, and their students' parents.

In most cases, the structural elements of schools that were related to rules, policies and hierarchical management systems gave the teachers unqualified permission to fully implement SSP through a socially-critical pedagogy. The teachers acknowledged that their principals had not only been responsible for the decision to implement SSP, but that they were also actively involved in supporting the teachers through the provision of time for professional development and curriculum design, the development of new and different learning spaces, and in some cases, direct mentoring of classroom practices.

Cathy for example, considered the incorporation of SSP into her teaching routine to be part of her usual process of trialing new approaches. She indicated that she felt supported in her efforts to constantly improve her practice, stating that if "I find something that works, okay I'll do it again, but, if I find something that doesn't work as well as I had hoped, then I'll look for an alternative". She believed that she was "allowed to do that" due to a "fairly progressive" principal who encouraged an "open minded approach" and "willingness" to "have a go". Cathy reported that her principal embraced the philosophy that "we've done it this way, why not try it another way, if it's successful, great…okay if you fail…we've learnt something out of it". Cathy's comment that "it's great to have an institution that can do that" suggested that her willingness to try different approaches may have been constrained if she had not had her principal's support. Similarly, David attributed the achievement of the SSP five-star accreditation at Ocean Primary School to the support of the school management, stating that "I'm really grateful…[be]cause from the boss down, and the school council, I have a lot of support…they've basically said yes to everything…I think it also helps because our boss understands…he's been quite willing to support us". David explained that, like Cathy, he considered his principal's attitude critical to the willingness of teachers to trial new ideas: "he's got quite an entrepreneurial approach…if you can think of how to do it then go for it sort of thing".

Elizabeth was the only teacher to report that she was implementing SSP activities without the direct support of her principal. She had developed SSP-related activities under the auspices of the school's previous principal, who had been not only responsible for introducing the program, but also extremely enthusiastic for the program's goals. Elizabeth's current principal was unaware of SSP and unsure of its status in the school, despite the fact that the school continued to advertise their SSP five-star accreditation. Although Elizabeth maintained her waste-management routine, she reported that other teachers had stopped implementing SSP programs, stating that, for example, during the "year before last…we had a fairly big energy-wise education program, and we had an energy monitor for each room, and that person was in charge of turning off all the switches each night…this year we haven't".

However, irrespective of the level of support received from a principal, many of the teachers did not implement SSP through a socially-critical pedagogy. Lisa for example, employed a strong vocational/neo-classical pedagogy (see Fig. 5.5) despite unequivocal support from Helen. At other schools, a principal's strong support for SSP both enabled and constrained the teachers' implementation of SSP. This was particularly true at East Valley Primary School.

Several of the teachers at East Valley Primary School provided valuable insights into the potential for a principal's support to both enable and constrain a teacher's ability to implement a specific pedagogy. Karen for example, commented that her principal strongly supported all of her efforts to facilitate learning experiences at the East Valley Nature Park (EVNP) for all students. However, as Karen's classroom was removed from the school campus, and because many of the school's teachers did not visit EVNP, she did not believe that her work was unduly influenced by the actions or beliefs of her teacher colleagues. This was not the case for the teachers who taught in classrooms at the main East Valley Primary School campus. The enabling effect of a principal's assistance in terms of mentoring in classroom technique was best highlighted by Julia and Anita, who both attributed their ability to move away from their usual vocational/neo-classical practices to their principal's expertise because "we were floundering without her" (Julia). Julia reported that her principal "trains us all" and "is guiding us through the process of designing our curriculum". The liberal-progressive aspects of their pedagogy (see Fig. 5.1) reflected their attempts to incorporate an "integrated enquiry" approach to their teaching, which they described as "something very new to us" (Julia). However, although teachers such as Julia and Anita found their principal's support very useful, they also commented that it created tensions within the school. Their willingness to be supported in this manner was seen as a betrayal by colleagues who resisted change, resented SSP as an imposition on their well-established routines, and were "snobby" (Anita) in their dismissal of any practice that did not correlate with a vocational/neo-classical pedagogy. Robyn, Julia and Anita each noted that they were "very lucky because we get together and plan" (Anita) but that they could not "talk with, say, the other Prep teachers" (Anita) due to these tensions in their school. They commented that on any given day, the teachers in adjacent classrooms may not speak with each other, and would not know what activities each were undertaking with their students. Robyn noted that "other classes [are] sort of very much, very, segregated. We're doing our own thing. Someone else is doing theirs…[there is] not much talk about what each of us is doing".

Robyn, Julia and Anita each commented that the tensions in the school made their working environment uncomfortable, and reduced their confidence to successfully undertake the changes that their principal supported. The tensions at East Valley Primary School described by these teachers highlighted the potential for low teacher morale, or reduced ontological security, to thwart any change process (Gitlin and Margonis 1995; Hargreaves and Fullan 1998), a notion supported by the observation that "educational change is hard to implement effectively because it is often resented and resisted, and because it often creates dissatisfaction, lowered morale and demotivation" (L. Evans 2000, p. 188). The resistance of some of the teachers to change at East Valley Primary School influenced the morale and motivation of those who supported change, and indicated that the teachers' practices were not only influenced by the presence or absence of permission and practical support from their principals, but also from their colleagues.

However, the role of the principal in the formation of tensions at East Valley Primary School was not clear. The factors that influence the ability of a principal to successfully bring about change in their school have been the subject of much research (e.g. Davis et al. 2005; Mitchell and Castle 2005; Spillane 2006; Timperley 2005) but remain poorly understood (Gaussel 2007). Gaussel (2007) noted that "researchers do, however, agree on the fact that the influence of the principal has more to do with his/her personality than with his/her effective institutional power" (p. 3). In terms of the ideals of structuration, and the ontology-in-situ framework, Gaussel's comments indicate that the teachers' perceptions of permission to implement SSP may have influenced their ability to do so less than their perception of the personal qualities of, or their personal relationship with, their principal. The influence of the principals on the ability of the teachers to embrace the pedagogical changes demanded by the implementation of SSP is not discussed in detail here; none-the-less, it was apparent that the principals were aware that their actions did influence the teachers' practices. In particular, the principals were most concerned about the potential of their public support for those teachers who agreed to implement the program to create tension in their schools.

It is important to note that the principals from all but two schools were unwilling to formally discuss issues related to the implementation of SSP, despite their acknowledgement that they were responsible for the decision to implement the program, and that they desired SSP five-star accreditation in order to advertise their school. Several principals voiced their concern for the potential detrimental effects that public discussions could have on their efforts to implement SSP in their school. These principals acknowledged the difficulties they faced in mandating pedagogical change, and the tensions this created amongst their teachers. Some principals acknowledged that they deliberately focussed their support towards those teachers willing to engage with SSP in the hope that others would, in time, see the benefits and feel less threatened by the changes.

In addition to acknowledging the need to gain the support of school principals and colleagues in order to feel comfortable to trial new practices, the teachers indicated that the opinions and support of their students' parents was also valued. David, for example, noted the initial difficulty of implementing "studies of Asia" at Ocean Primary School: "people [the parents] were suspicious" and told the school that "I'll never need that and why can't they [the students] learn this [something else]". David stated that "there were a lot of things to get over". He explained that the teachers had felt extremely uncomfortable at having to justify the new studies of Asia to the parents. They explained that:

> Asia…they're our neighbours, why wouldn't we be doing studies of Asia…it doesn't mean we're going to change everybody's philosophy or religion or eating habits you know, we can just appreciate that some people look differently, dress differently and bring different things to lunch.

However, he noted that even though "studies of Asia is probably the least integrated [aspect of the curriculum] so far…people, including parents, don't question [its place in the school curriculum] any more". He noted that it took several years for the parents to accept that "studies of Asia isn't a subject, [just as] the environment isn't a subject…they are themes" to be incorporated and integrated into many aspects of learning.

6.5.2   Knowledge Required to Implement SSP

Educational researchers have previously attributed different types of teacher 'knowledge' to the development of rhetoric–reality gaps. For example, Schweisfurth (2006) related educational rhetoric–reality gaps to "space" in a curriculum (p. 210). 'Space' in any curriculum may only be effectively utilised to achieve specific goals by the teachers with the understanding, or knowledge, to do so. Schweisfurth (2006) noted that specific educational goals, such as the incorporation of global issues, "may not be explicit" in a curriculum, but that a curriculum may instead provide "the space for teachers" to engage with such goals. In other words, the structural elements of a curriculum may require teachers to "engage in issues they feel are important" (p. 210). Although this example highlights the role of agency in teachers' practices, it does not adequately explain the rhetoric–reality gaps in the implementation of SSP. Researchers have also attributed ineffective environmental education to inadequate teacher knowledge regarding environmental concepts (e.g. Cutter and Smith 2001a; Said et al. 2003; Spork 1992), and this no doubt contributes to the development of many rhetoric–reality gaps. However, as identified by both the principals and the teachers, SSP was not a typical curriculum. SSP documents incorporated specific instructions for a sequence of mandatory activities to be undertaken by both the schools and the teachers in order to implement the program and to obtain their SSP five-star accreditation (refer to Tables 2.1 and 2.2). These were designed to assist school communities to begin to develop a deeper understanding of SSP goals, irrespective of their initial environmental knowledge. In other words, implementing SSP through a socially-critical pedagogy positioned the teachers alongside their students as learners. Schools were also encouraged to modify or adapt other aspects of SSP implementation to suit their own interests and goals. Professional development sessions and collaborative work with the SSP facilitators from CERES assisted the schools and the teachers in these tasks.

More significantly however, the implementation of SSP was not predicated on the teaching or learning of specific environmental knowledge. All of the teachers demonstrated an excellent understanding of the rhetoric of SSP and correctly identified its future-oriented and social transformative goals of influencing human–environment relationships. In addition, the teachers identified that SSP goals were best achieved through the development of students' attitudes, sense of responsibility, curiosity and confidence to question, that is, through the development of aspects of students' agency. In other words, the lack of teacher knowledge, regarding either specific environmental knowledge or understanding of the goals or requirements of SSP, did not adequately explain the rhetoric–reality gaps in the implementation of the program in the classroom. It was therefore important to consider the teachers' knowledge of the pedagogical practices required to achieve the SSP goals that they identified.

6.5.3   Implementing a Socially-Critical Pedagogy

The need to implement a socially-critical pedagogy, as advocated by SSP (and discussed in Sect. 2.3), has been considered responsible, in part, for the prevalence of rhetoric–reality gaps related to environmental education (e.g. Chapman 2004; Oulton and Scott 2000; Robinson 1994; Scott and Oulton 1999; Walker 1997). Researchers have reported that many teachers find a socially-critical pedagogy difficult to embrace because it "fails to give them an implementation theory" (Walker 1997, p. 161), and because "schools are structured in such a way that they cannot accommodate the radical social change required" (Robinson 1994, p. 60). In addition, implementing a socially-critical pedagogy forces teachers to "reconceptualise their curriculum and to question prevailing practices" (Walker 1997, p. 158). The implementation of a socially-critical pedagogy is a complex and multifaceted process. Despite this, the implementation of a socially-critical pedagogy was not only a major motivation for the decision by principals to establish SSP within their schools, but was also recognised by the teachers as essential for achieving the goals of SSP.

6.5.4   Teacher Experience

The stories in Chap. 5 indicated that the teachers who were implementing SSP included both recent graduates and those who had been teaching for over twenty years. Similarly, the teachers worked at schools that were either just beginning to implement SSP, or that had been implementing the program for at least five years. A reasonable assumption may have been that the teachers most willing to embrace pedagogical change in order to implement SSP might have been the younger, albeit less-experienced teachers—those who had less well-established classroom routines and who may have experienced a range of pedagogies during their own schooling—but this was not observed. Comparison of the pedagogy employed by the teachers to implement SSP with their years of teaching showed that, although the less-experienced teachers were less likely to have introduced a socially-critical pedagogy, level of experience in terms of either years of teaching, or years of implementing SSP, did not predict a teacher's ability to implement a socially-critical pedagogy.

Observations of novice teachers by several researchers (e.g. Grossman 1990; Korthagen 2001; Miles and Cutter-Mackenzie 2006) suggest that, as noted here, the less-experienced teachers are not always able to act as agents of change. Developing and establishing classroom practices are demanding tasks that often led the novice teachers to conform to the dominant practices in a school rather than work to introduce new pedagogies they may, or may not, have practiced during their training (Korthagen 2001). The pressure to conform to the established practices within a school work environment was highlighted by Julia, who upon joining the school, found herself conforming to a pedagogy she recognised as outdated and inadequate.

She described staff planning meetings organised by the more-experienced teachers: "week one, let's do the life cycle of frogs, week two, let's do the life cycle of chickens, every week we'd have the worksheets…all the grades did the same thing each week".

However, that is not to say that the less-experienced teachers did not contribute in a valuable way to the implementation of SSP. Helen noted that the younger, less-experienced teachers were most enthusiastic about SSP and excited by the opportunity to develop and share new ideas. She reported that the more-experienced teachers were better able to find ways in which to most effectively put these new ideas into practice. Helen suggested that the best way in which to implement SSP throughout a school was to establish a collaborative work environment in which teachers with different levels of experience worked together: a notion not completely dissimilar to the benefits of multi-age learning activities of a socially-critical pedagogy. The idea that successful educational change is best achieved through a strategy of teacher collaboration is not new (e.g. Carson 2007; Fullan 2001; Greenfield 2005), and reflects the notion that work place culture, not educational policy, determines the effectiveness of any change process (Carson 2007). Greenfield (2005) suggested that:

> one of the biggest challenges that successful leadership in schools entails [is] to encourage and support collaboration among teachers that results in improved teaching practices and desired learning outcomes for children, that is, to develop the school as a community of professionals working together to serve children well (p. 246).

6.6 Rhetoric, Reality and Educational Rhetoric–Reality Gaps

Analysis of the rhetoric represented in both the teachers' and principals' stories (see Chap. 5) indicated that these educators had a good understanding of the future-oriented and socially transformative outcomes of SSP, and agreed that these fulfilled the democratic purposes of education. Similarly, both the principals and the teachers had a good understanding of the differences between a vocational/neo-classical pedagogy and the socially-critical pedagogy advocated by SSP. However, one significant difference between the principals and the teachers related to their motivation for implementing SSP. The teachers implemented SSP in response to their principal's directions, with the view that the program would enable them to achieve the socially transformative educational outcomes they identified in SSP documents. The principals on the other hand, stated that their decision to implement SSP was based not solely on its educational outcomes, but for its potential as a vehicle for pedagogical change. Irrespective of these divergent motivations, the teachers believed that they had permission to implement SSP, and that a socially-critical pedagogy was the best way in which to achieve the educational outcomes of SSP.

SSP documents provided a detailed sequence of activities and goals to assist the schools and the teachers to fully and effectively implement the program (see Chap.

2). The teachers were also tutored and attended a variety of professional development sessions, many of which were undertaken in their own classrooms. In other words, not only did the teachers have permission to implement SSP, they also received significant professional and personal assistance to introduce the necessary socially-critical pedagogy. Despite that, all but two of the teachers chose to implement SSP with a pedagogy other than the socially-critical approach advocated by SSP and their principals (see Figs. 5.1, 5.2, 5.3, 5.4, 5.5 and 5.6).

The teachers' rhetoric indicated that a teacher's knowledge of the educational outcomes of SSP and of the characteristics of a socially-critical pedagogy did not predict the presence, or absence, of a rhetoric–reality gap in their attempts to implement SSP. Similarly, the presence of a rhetoric–reality gap in a teacher's practice of SSP did not predict either the lack, or availability, of professional assistance and training, a principal's permission, or collegial support for that teacher. A comparison of the pedagogy used by teachers and their years of experience indicated that although the less-experienced teachers were less likely to have introduced a socially-critical pedagogy, the level of experience in terms of either years of teaching, or years of implementing SSP, did not indicate whether or not a teacher would implement a socially-critical pedagogy, and therefore, did not predict the likelihood of a rhetoric–reality gap. These factors therefore represent aspects of the implementation of SSP which did not significantly contribute to the rhetoric–reality gaps identified.

Thus, in order to identify the factors (ontological elements) that most contributed to the development of educational rhetoric–reality gaps it was essential to identify those aspects of the implementation of SSP that significantly enabled and/or constrained the teachers' practices. The teachers' stories (see Chap. 5) indicated that the teachers considered the specific practicalities of undertaking a socially-critical pedagogy to have most strongly influenced their ability to implement SSP, including: access to suitable learning spaces; the effects of routine and time; and the availability of other teaching and learning resources. The role of each of these in the development of educational rhetoric–reality gaps, represented by the vocational/neo-classical and liberal-progressive practices of some of the teachers, is discussed in Chap. 7.

References

Beare, H. (2001). *Creating the future school*. London: Routledge.

Burbules, N. C., & Berk, R. (1999). Critical thinking and critical pedagogies: Relations, differences, and limits. In S. P. Thomas & L. Fendler (Eds.), *Changing terrains of knowledge and politics* (pp. 45–65). New York: Routledge.

Carneiro, P., Crawford, C., & Goodman, A. (2006). *Which skills matter?* London: Centre for the Economics of Education, University of London.

Carson, J. A. (2007). *The ecology of school change: An Australian primary school's endeavor to integrate concept-based, experiential environmental learning throughout core curriculum*. PhD, The University of Arizona, USA.

Chapman, D. (2004). Sustainability and our cultural myths. *Canadian Journal of Environmental Education, 9*, 92–108.

Cotton, D. R. E. (2006). Implementing curriculum guidance on environmental education: The importance of teachers' beliefs. *Journal of Curriculum Studies, 38*(1), 67–83.

Cranston, N., Mulford, B., Keating, J., & Reid, A. (2009, July 2–4). *Researching the public purposes of education in Australia: the results of a national survey of primary school principals.* Paper presented at the Australian Association for Research in Education Conference, Canberra, Australia. http://www.aare.edu.au/publications-database.php/5850/Researching-the-public-purposes-of-education-in-Australia:-The-results-of-a-national-survey-of-primary-school-principals. Accessed 1 Oct 2014.

Cranston, N., Kimber, M., Mulford, B., Reid, A., & Keating, J. (2010). Politics and school education in Australia: A case of shifting purposes. *Journal of Educational Administration, 48*(2), 182–195.

Cutter, A., & Smith, R. (2001a). A chasm in environmental education: What primary school teachers 'might' or 'might not' know. In B. Knight & L. Rowan (Eds.), *Researching in contemporary educational environments* (pp. 113–132). Flaxton: Post Pressed.

Cutter, A., & Smith, R. (2001b). Gauging primary school teachers' environmental literacy: An issue of priority. *Asia Pacific Education Review, 2*(2), 45–60.

Davis, S., Darling-Hammond, L., LaPointe, M., & Meyerson, D. (2005). *School leadership study: Developing successful principals.* Stanford: Educational Leadership Institute.

DEST. (2005). *National framework for values education in Australian schools.* Department of Education, Science and Training (DEST), Canberra. http://www.curriculum.edu.au/verve/_resources/Framework_PDF_version_for_the_web.pdf. Accessed 30 Sept 2014.

Doll, J., & Ajzen, I. (1992). Accessibility and stability of predictors in the theory of planned behavior. *Journal of Personality and Social Psychology, 63*(5), 754–765.

Evans, R. (1996). *The human side of school change.* San Francisco: Jossey-Bass.

Evans, L. (2000). The effects of educational change on morale, job satisfaction and motivation. *Journal of Educational Change, 1*, 173–192.

Feather, N. T. (1992). Values, valences, expectations and actions. *Journal of Social Issues, 48*(2), 109–124.

Feinstein, L. (2000). *The relative economic importance of academic, psychological and behavioural attributes developed in childhood.* London: University of London.

Fien, J. (1993). *Education for the environment: Critical curriculum theorising and environmental education.* Geelong: Deakin University Press.

Fien, J. (2001). *Education for sustainability: Reorientating Australian schools for a sustainable future.* Tela series, Australian Conservation Foundation, Fitzroy, Australia. http://www.acfonline.org.au/sites/default/files/resources/tela08_education_%20for_sustainability.pdf. Accessed 30 Sept 2014.

Fishbein, M., & Ajzen, I. (1975). *Belief, attitude, intention, and behavior. An introduction to theory and research.* Reading: Addison-Wesley Publishing.

Fullan, M. (2001). *The new meaning of educational change* (3rd ed.). New York: Teachers College Press.

Gaussel, M. (2007). *Leadership and educational change.* Lettre d'information de la VST, 24. http://ife.ens-lyon.fr/vst/LettreVST/english/24-january-2007_en.php. Accessed 1 Oct 2014.

Gayford, C. (1996). The nature and purpose of environmental education. In G. Harris & C. Blackwell (Eds.), *Environmental issues in education* (pp. 1–20). Hants: Ashgate Publishing.

Giddens, A. (1979). *Central problems in social theory: Action, structure and contradiction in social analysis.* London: Macmillan.

Giddens, A. (1984). *The constitution of society.* Cambridge: Polity Press.

Gitlin, A., & Margonis, F. (1995). The political aspects of reform: Teacher resistance as good sense. *American Journal of Education, 103*(4), 377–405.

Gough, A. (1997). *Education and the environment: Policy, trends and the problems of marginalisation.* Melbourne: Australian Council for Educational Research.

Grace, M., & Sharp, J. (2000). Exploring the actual and potential rhetoric-reality gaps in environmental education and their implications for pre-service teacher training. *Environmental Education Research, 6*(4), 331–345.

Greenfield, W. D. (2005). Leading the teacher work group. In L. W. Hughes (Ed.), *Current issues in school leadership* (pp. 245–264). Mahwah: Lawrence Erlbaum.

Grossman, P. L. (1990). *The making of a teacher: Teacher knowledge and teacher education.* New York: Teachers College Press.

Gynnild, V. (2002). Agency and structure in engineering education: Perspectives on educational change in light of Anthony Giddens' structuration theory. *European Journal of Engineering Education, 27*(3), 297–303.

Hargreaves, A., & Fullan, M. (1998). *What's worth fighting for out there?* New York: Teachers College Press.

Huckle, J. (1986). Ten red questions to ask a green teacher. *Education Links, 37,* 4–8.

Jickling, B., & Spork, H. (1998). Education for the environment: A critique. *Environmental Education Research, 4*(3), 309–328.

Kelly, T. M. (1986). Discussing controversial issues: Four perspectives on the teacher's role. *Theory and Research in Social Education, 14*(2), 309–327.

Kemmis, S., Cole, P., & Suggett, D. (1983). *Orientations to curriculum and transition: Towards the socially-critical school.* Melbourne: Victorian Institute of Secondary Education.

Korthagen, F. (2001). *Linking practice with theory: The pedagogy of realistic teacher education.* Mahwah: Lawrence Erlbaum.

Labaree, D. F. (1997). Public goods, private goods: The American struggle over educational goals. *American Educational Research Journal, 34*(1), 39–81.

Margo, J., Dixon, M., Pearce, N., & Reed, H. (2006). *Freedom's orphans: raising youth in a changing world* (Vol. January 31). London: Institute for Public Policy Research.

MCEETYA. (2008). *Melbourne declaration on educational goals for young Australians.* Ministerial Council on Education, Employment, Training and Youth Affairs, Australia. http://www.curriculum.edu.au/verve/_resources/National_Declaration_on_the_Educational_Goals_for_Young_Australians.pdf. Accessed 1 Oct 2014.

McRobb, S., Jefferies, P., & Stahl, B. C. (2007). Exploring the relationship between pedagogy, ethics and technology: Building a framework for strategy development. *Technology, Pedagogy and Education, 16*(1), 111–126.

Miles, R., & Cutter-Mackenzie, A. (2006). *Environmental education: Is it really a priority in teacher education.* Paper presented at the Sharing wisdom for our future: environmental education in action.

Mitchell, C., & Castle, J. B. (2005). The instructional role of elementary school principals. *Canadian Journal of Education, 28*(3), 409–433.

Mogensen, F. (1997). Critical thinking: A central element in developing action competence in health and environmental education. *Health Education Research, 12*(4), 429–436.

Mulford, B., & Edmunds, B. (2010). *Educational investment in Australian schooling: Serving public purposes in Tasmanian primary schools.* Hobart: University of Tasmania.

O'Connell, R. F., Ely, M., Krasnow, M., & Miller, L. (2001). Professional development of change agents: Swimming with and against the currents. In R. F. O'Connell (Ed.), *Guiding school change: The role and work of change agents* (pp. 16–36). New York: Teachers College Press.

Oulton, C., & Scott, W. (2000). A time for revisioning. In B. Moon, M. Ben-Peretz, & S. Brown (Eds.), *Routledge companion to education* (pp. 489–501). London: Routledge.

Raulo, M. (2000). Moral education and development. *Journal of Social Philosophy, 31*(4), 507–518.

Robertson, C. L., & Krugly-Smolska, E. (1997). Gaps between advocated practices and teaching realities in environmental education. *Environmental Education Research, 3*(3), 311–326.

Robinson, V. M. J. (1994). The practical promise of critical research in educational administration. *Educational Administrative Quarterly, 30*(1), 56–76.

Robinson, V. M. J. (2006). Putting education back into educational leadership. *Leading and Managing, 12*(1), 62–75.

Robinson, V. M. J. (2007, August 12–14). The impact of leadership on student outcomes: making sense of the evidence. In *Australian Council for Educational Research Conference, The leadership challenge: improving learning in schools, Melbourne, Australia* (pp. 12–17).

Rohan, M. J. (2000). A rose by any name? The values construct. *Personality and Social Psychology Review, 4*(3), 255–277.

Sahid, A. (2004). The changing nature of the role of principals in primary and junior secondary schools in South Australia following the introduction of local school management. *International Education Journal, 4*(4), 144–153.

Said, A. M., Ahmadun, F., Paim, L. H., & Masud, J. (2003). Environmental concerns, knowledge and practices gap among Malaysian teachers. *International Journal of Sustainability in Higher Education, 4*(4), 305–313.

Schofield, K. (1999). *The purposes of education 3. A contribution to the discussion on 2010: Final report.* Brisbane: Queensland State Education.

Schweisfurth, M. (2006). Education for global citizenship: Teacher agency and curricular structure in Ontario schools. *Educational Leadership, 58*(1), 41–50.

Scott, W., & Gough, S. (2003). *Sustainable development and learning: Framing the issues.* London: Routledge.

Scott, W., & Gough, S. (2004). *Key issues in sustainable development and learning: A critical review.* London: Routledge.

Scott, W., & Oulton, C. (1999). Environmental education: Arguing the case for multiple approaches. *Educational Studies, 25*(1), 89–97.

Spillane, J. P. (2006). *Distributed leadership.* San Francisco: Josey-Bass.

Spork, H. (1992). Environmental education: A mismatch between theory and practice. *Australian Journal of Environmental Education, 8,* 147–166.

Stevenson, R. B. (1987). Schooling and environmental education: Contradictions in purpose and practice. In I. Robottom (Ed.), *Environmental education: Practice and possibility* (pp. 69–82). Burwood: Deakin University Press.

Stevenson, R. B. (2007). Editorial. *Environmental Education Research, 13*(2), 129–138.

Timperley, H. S. (2005). Distributed leadership: Developing theory from practice. *Journal of Curriculum Studies, 37*(4), 395–420.

Tomlins, B., & Froud, K. (1994). *Environmental education: Teaching approaches and students' attitudes: A briefing paper.* Slough: UK Foundation for Educational Research.

UNESCO. (2005). *United Nations decade of education for sustainable development 2005–2014. International implementation scheme.* United Nations Educational, Scientific and Cultural Organisation. http://unesdoc.unesco.org/images/0014/001486/148654e.pdf. Accessed 1 Oct 2014.

UNESCO. (2007, November 26–28). *Moving forward from Ahmedabad. Environmental education in the 21st century.* Fourth Intergovernmental Conference on Environmental Education, Ahmedabad (India), United Nations Educational, Scientific and Cultural Organisation. http://www.unevoc.net/fileadmin/user_upload/docs/AhmedabadFinalRecommendations.pdf. Accessed 1 Oct 2014.

Vare, P., & Scott, W. (2007). Learning for a change: Exploring the relationship between education and sustainable development. *Journal of Education for Sustainable Development, 1*(2), 191–198.

Walker, K. (1997). Challenging critical theory in environmental education. *Environmental Education Research, 3*(2), 155–162.

Watson, L. (2009). Issus in reinventing school leadership: Reviewing the OECD report on improving school leadership from an Australian perspective. *Leading and Managing, 15*(1), 1–13.

Wink, J. (2000). *Critical pedagogy: Notes from the real world* (2nd ed.). New York: Addison Wesley Longman.

Chapter 7
The Dance of the 'Duality of Structure and Agency'

The implementation of the Sustainable Schools Program (SSP) was accompanied by the development of educational rhetoric–reality gaps. Such gaps were represented by the incongruence of a teacher's classroom pedagogy and self-description of that pedagogy, and between a teacher's understanding of the rhetoric of SSP and actual implementation of SSP. Most significantly, when asked to implement SSP through the mandated socially-critical pedagogy, most teachers failed to do so, and chose a vocational/neo-classical or liberal-progressive approach. The case studies of the teachers who were required to implement SSP indicated that the practicalities of undertaking a socially-critical pedagogy most strongly influenced a teachers' ability to effectively implement the program. This chapter draws on Giddens' theory of structuration to identify the critical ontological elements of structure and agency that both constrained and enabled the teachers to deal successfully with the practicalities of implementing a socially-critical pedagogy, most particularly in relation to: learning spaces (Sect. 7.1); routine and time (Sect. 7.2); and other learning resources (Sect. 7.3). Giddens' notion of the duality of structure and agency informs the understanding that relationships between these ontological elements defined the major differences between the teachers whose practices represented best practice, and those whose practices represented a rhetoric–reality gap in the implementation of SSP through a socially-critical pedagogy.

7.1 Learning Spaces

A socially-critical pedagogy requires both the students and the teachers to not only re-define their roles in the learning process, but to also re-define what constitutes a learning space. The rhetoric–reality gaps in the implementation of the Sustainable Schools Program (SSP) indicated that some teachers were unable to re-define their practices in these ways. These teachers' experiences provided valuable insights into

© Springer International Publishing Switzerland 2016
J. Edwards, *Socially-critical Environmental Education in Primary Classrooms*,
International Explorations in Outdoor and Environmental Education 1,
DOI 10.1007/978-3-319-02147-8_7

the ways in which learning spaces both shaped, and were shaped by, their practices. Although all learning environments incorporate a range of both social and physical aspects, the term learning space is used here to refer to just the physical attributes of a learning environment.

The notion of learning spaces was a prominent theme in the SSP documents and the rhetoric of both the principals and the teachers. Indeed, the SSP five-star accreditation process demanded that schools make significant changes to aspects of the management, organisation and design of both indoor and outdoor learning spaces. This was seen to position education within a more sustainable learning environment, and reflected an understanding that if a socially-critical pedagogy was to become widely and effectively implemented, the design of new educational learning spaces must incorporate:

> an awareness of the need for diverse types of learning spaces to offer multiple approaches to the acquisition of different sorts of knowledge or skills, and a greater emphasis on environments that recognise learner–learner interactions as well as learner–teacher interactions (Rudd et al. 2006, p. 9).

The development of new learning spaces was often presented as important evidence of a school's progression towards the effective implementation of SSP, and considered a necessary resource for motivating and enabling the teachers to undertake a socially-critical pedagogy. It is interesting to note that the schools were provided with a step-by-step process for effectively transforming and/or developing learning spaces in order to achieve SSP five-star accreditation, but no similar process was provided to guide the teachers in how to transform their pedagogy in order to more effectively use these new learning spaces.

The notion that learning spaces are critical to achieving particular educational outcomes is not a new idea. For example, Lippman (2002) argued that traditional classrooms represent learning spaces designed to accommodate the "short term information mastery goals" of a traditional vocational/neo-classical pedagogy, characterised by "a single adult interacting with many in relative impersonal social relations in which social rules, principles, and guidelines govern the activity" (p. 5). Similarly, Van Note Chism (2006) noted that "traditional classrooms tend to be designed on the basis of transmission theory whose built pedagogy says that one person will 'transfer' information to others who will 'take it in' at the same rate by focusing on the person at the front of the room" (quoted in Rudd et al. 2006, p. 9). Lippman (2002) believed that such learning spaces were designed primarily to "control behaviour" (p. 5), with the effect that they "reinforce for children that they have little power to make changes in their daily lives, affect their environment, or [have] opportunities to examine alternative ways of living" (p. 5). As discussed earlier, in light of these ideas, and the notion that every space is a learning space, the principals often justified their decision to implement SSP according to the need to develop learning spaces as vehicles for change (e.g. the new school buildings at South Bay Primary School), or to better use existing spaces (e.g. the outdoor spaces at Ocean Primary School).

7.1.1 *Learning Spaces: Vehicles for Change*

At some of the schools, the impending building of new facilities was an important factor in the decision to implement SSP, and reflected the principals' beliefs that new learning spaces would motivate the teachers to embrace pedagogical change. Several of the teachers indicated that the provision of new and different learning spaces was absolutely essential to their ability to implement certain types of pedagogy and to provide different learning experiences.

Lisa, for example, commented that the layout and design of the current school buildings and classrooms made it very difficult for her to alter her existing pedagogy. In particular, she noted the necessity to be constantly "moving rooms" or "moving furniture" in order to accommodate the activities she believed to be best suited to a socially-critical pedagogy. She hoped that the new classrooms would enable her to "accommodate" opportunities for shared learning through the interaction of students in different classes. Lisa was also adamant that she was unable to change her well-established vocational/neo-classical pedagogy to a socially-critical pedagogy until she had access to what she considered to be an "ideal classroom". She described such a learning space as still needing "four walls" but which also provided "access straight out into outdoor learning areas", because "I'd love to have an area where you could work outdoors". Barrett (2007) notes that the "ability to take students outside" is commonly "cited as a problem" by teachers when questioned about their inability to implement effective environmental education. However, as Lisa's school had recently completed the development of a range of outdoor facilities, including a frog pond, and native and vegetable gardens, and was situated near a variety of community and natural spaces, Lisa's notion of what was required to work outdoors was not easy to determine. Her comment that she could not expect the students to sit outside in the "direct sun" may have indicated concerns regarding health and safety, but this was not supported by a previous decision to allow the students to walk along the local river for a water quality project. Alternatively, Lisa's comments suggested that she was searching for a way in which to merely transfer her existing classroom practices into an outdoor setting rather than implement more participatory or socially-critical approaches.

Lisa's case highlighted a disconnect between the principals' rhetoric regarding the need to provide learning spaces as motivation for pedagogical change, and the teachers' references to the lack of appropriate learning spaces as justification for not being able to implement pedagogical change. In Giddens' terms, Lisa's access to a range of new and different learning spaces suggested that her rhetoric concerning her inability to undertake change reflected a discursive consciousness (see Fig. 4.2), that is, a verbal justification that reflected underlying values that prevented Lisa from implementing a socially-critical pedagogy, not an actual lack of learning spaces.

Like Lisa, Elizabeth indicated that Mountain Primary School had developed a wide range of learning spaces. She proudly explained how important the kitchen and native gardens with shaded courtyards and outdoor seating were in demonstrating to the local community Mountain Primary School's ability to implement SSP. However, her use of these facilities, along with the multitude of easily accessible community and outdoor learning spaces close to the school, was limited to augmenting the knowledge acquisition component of pre-established curriculum projects (see Fig. 5.2). Although Elizabeth did indicate that some aspects of the learning spaces she used would benefit from better design, for example, allowing her students to share work with other classes depended on "if it's quick and easy to get to that room you can do it", unlike Lisa, she did not equate her use of a vocational/neo-classical to inadequate facilities.

The principals at other schools justified the implementation of SSP as a way of making better use of existing outdoor learning spaces. At Ocean Primary School, for example, David explained that because the school was situated on "a very large site" and because all of this land represented "a learning area...[as] a learning area isn't just a classroom" the principal sought to address parental and social expectations that the school be "environmentally responsible" with this land. Despite ready access to outdoor learning spaces, David explained that in order to effectively use these "special learning areas" specific, appropriate facilities were required to be developed. He indicated that, through the implementation of SSP, such outdoor facilities had been developed. He explained:

> my kids just love it when I take them out to the farm and to see those things growing...it's different to, you know, the old equivalent thing was the little saucer of cotton with the little seed growing out of [it], well now we've got a hot house out there, now the Preps [preparatory year students] can have their own vegetable garden...so all those things have added to help it [acceptance of SSP]...material changes which have added to that momentum [for change].

David believed that the provision of specific learning facilities, such as an "indigenous garden", had legitimised the use of Ocean Primary School's land for outdoor learning. As demonstrated by David's use of a new courtyard as a convenient site for his students to investigate issues related to human behaviour and the creation and disposal of litter, David viewed these outdoor learning facilities as "nice" environments, provided and designed by the school, within which learning tasks could be undertaken. Students' contribution to the planning, design, and development of these outdoor areas was limited. David noted that the improvements to the outdoor areas were considered by many of the teachers to be the school's ultimate goal for introducing SSP, that is, to "provide a nice learning environment" rather than to facilitate the continuing development of new pedagogies. He stated that "now what I want to address in the future...is complacency", because many of the teachers held the attitude that "oh sure the [courtyard] looks nice so now we don't have to do anything more".

Thus, although the implementation of SSP provided a reason, and momentum, for Ocean Primary School to improve many aspects of their existing outdoor learning spaces, the potential for these learning spaces to contribute to pedagogical change was not fully realised. In other words, and as demonstrated by both Lisa and Elizabeth, the provision of a range of learning spaces with the physical or structural features conducive to SSP-related activities did not guarantee that the teachers could, or would, implement a socially-critical pedagogy.

7.1.2 Every Space a Learning Space

The teachers who claimed to have inadequate spaces for learning, such as Lisa, often made little effort to alter either their teaching environment or their teaching practices so that they could more effectively utilise the spaces around them. The teachers who most successfully utilised a range of learning spaces, irrespective of the age or design of their classroom and school facilities, made conscious and deliberate efforts to either adapt each learning space to their students' needs, or to adapt their pedagogical approach to make the most of the learning space at hand. In other words, the ability to use any space as a learning space depended on the teachers' agency.

Both Cathy and Karen effectively utilised a range of learning spaces in order to implement SSP through a socially-critical pedagogy (see Figs. 5.1 and 5.4). Cathy enjoyed a school environment which, although not extensive, was well-designed to incorporate areas of native gardens and natural bush land. Although Cathy's approach to SSP incorporated projects which focused on the use of these outdoor areas, as well as a variety of learning spaces outside the school grounds, the majority of her work was based in a relatively small and traditional classroom equipped with the usual array of student furniture, book cases, cupboards and a white board. Unlike Lisa, the structural constraints of a traditional classroom learning space did not constrain her ability to implement SSP, to the extent that her practices exemplified a socially-critical pedagogy. Cathy understood that any space could become a space for learning if it met the needs of the students: a belief supported by the notion that learning space "is first and foremost about education, not architecture" (Rudd et al. 2006, p. 3).

Cathy used a socially-critical pedagogy to facilitate student–student interactions in her traditional classroom, by encouraging the students to negotiate, collaborate and cooperate in organising the learning space in any way that met their needs. As a result, Cathy's classroom represented a constantly changing learning space quite unlike the static and uncompromising setup of Lisa's classroom. Cathy actively invited the students to identify potential learning spaces and to find ways in which to utilise them, stating that "what we've found [is that] things that have sort of cropped up since we started [SSP] have been fantastic programs, for instance, our nesting box program…initiated by one of the year 4 girls", which involved a scientific study of birds in a previously unused area of bush land along the school boundary. In contrast, Lisa noted that the students were not encouraged to participate in the development of learning spaces at her school, stating that, for example: "we've got some veggie gardens up the back now, but I don't know where that idea came from, it just seemed to appear one day…I think that it was a parent [who] did it…I don't really know where the veggie garden came from". Cathy demonstrated that the most effective development of learning spaces occurred when the teachers enabled the students to negotiate and cooperate in the development of those spaces— a central element of effective socially-critical pedagogy and SSP (see Chap. 2).

On the other hand, Karen taught within the most unique learning spaces of all of the schools: a classroom within the grounds of the East Valley Nature Park (EVNP). Although Karen's expertise was in Information and Communication Technology (ICT) education, she incorporated a wide range of exciting and unusual outdoor and real life learning experiences which she made possible by utilising all of the resources of the nature park. These resources included the 'experts', or park staff, who were able to assist with the design and implementation of appropriate and authentic learning experiences within all environments of the park. In other words, the diverse learning spaces of the nature park enabled Karen to implement a socially-critical pedagogy because Karen chose to make use of all the opportunities that those learning spaces offered.

The teachers' practices suggested that neither the provision, nor the lack, of learning spaces conducive to the requirements of SSP could predict the willingness or ability of a teacher to implement SSP through a socially-critical pedagogy. In addition, there was anecdotal evidence that certain learning spaces could significantly constrain a teacher's ability to do their work. This was most clearly indicated by Karen's reports of the teachers who accompanied their students to EVNP. Karen observed that most of her colleagues found these learning spaces to be "extremely threatening" due to "the fact that we even just walk out the front gate…the fact that we're here in this environment". She reported that those teachers were not only unable to cope with learning outside the school, but were also particularly concerned about the lack of facilities such as four-walled classrooms and bells to indicate lesson times. For those teachers, the learning spaces provided by EVNP constrained their ability to implement almost any pedagogy, not just the socially-critical pedagogy advocated by SSP. Karen, on the other hand, considered the well-established socially-critical pedagogical routine that she had developed at EVNP, her usual working environment, as "this is just an assumed part of our education". This supported the notion that the teachers defined their practices by the well-established routines they had developed in their most familiar learning spaces, and that the "teachers' fear of launching into the unknown" (Trautmann and MacKinster 2005, p. 1) often rendered such well-established teaching practices difficult to change.

However, at Mountain Primary School, the presence of some schoolyard facilities did encourage some of the teachers to move away from a strictly vocational/neo-classical pedagogy. For example, one teacher allowed Prep students to explore and test their newly developed mathematical skills by measuring such things as chicken legs and water weeds in the kitchen garden. The constant stream of questions from the students and the freedom they felt to interact with others as they moved around the learning space ensured that learning from this lesson was significantly broader than a single mathematical concept. The teacher commented that her use of the kitchen garden for this activity resulted from her observation that the students enjoyed the experience of learning in a different environment. The use of this learning space facilitated a degree of pedagogical diversity, as both the teacher and the students responded to their physical surroundings.

7.1.3 Learning Spaces, Teachers' Practices and Rhetoric–Reality Gaps

SSP was not intended to be undertaken only by those schools with extensive facilities or expansive grounds, or by those intending to re-build. SSP encouraged all school communities to work collaboratively with their students to not only identify potential new learning spaces, but most importantly, to also transform the way in which the teachers and the students interacted within any learning space. As identified by the teachers, the socially-critical pedagogy embraced by SSP required both the students and the teachers to not only re-define their roles in the learning process, but to also re-define what constituted a learning space. In light of this, it was reasonable to expect that rhetoric–reality gaps in the implementation of SSP indicated that the teachers were unable to re-define their practices in these ways. However, for these teachers, the role of learning spaces was not universally significant in the development of the rhetoric–reality gaps in the implementation of SSP.

The teachers' practices indicated that, contrary to the school principals' expectations, the provision of new and/or different types of learning spaces alone did not necessarily facilitate the implementation of a socially-critical pedagogy. Irrespective of the learning spaces available to a teacher, it was aspects of a teachers' agency that determined whether or not they successfully enacted a socially-critical pedagogy. Although learning spaces were important in assisting a teacher to practice in a particular way, they did not determine those practices. In addition, and in line with Giddens' notion of the duality of structure and agency, the teachers' practices influenced the design and/or utility of the learning spaces, irrespective of the teachers' preferred pedagogy. For example, the long rows of perfectly aligned desks in Lisa's classroom not only reflected her preference for students to remain silent and obediently attentive to her instruction at the front of the class, but also prevented students from interacting with each other, sharing ideas or working together in groups. This classroom was organised by Lisa to facilitate her vocational/neo-classical pedagogy, and as such, discouraged activities that fell outside that pedagogy. In contrast, the ever-changing layout of Cathy's classroom reflected the ideals of the socially-critical pedagogy advocated by SSP. Cathy's preference for a socially-critical pedagogy meant that she encouraged her students to actively participate in structuring their activities, and in so doing, to identify and develop learning spaces that addressed their needs. The teachers' use of learning spaces in the implementation of SSP suggested that neither the provision, nor the lack, of the type of learning spaces perceived to be conducive to the requirements of SSP could predict the willingness or ability of the teachers to implement the program through a socially-critical pedagogy. Thus, the rhetoric–reality gaps in the implementation of SSP could not be attributed to the learning spaces in which the teachers and the students were required to work. However, the rhetoric of the teachers suggested that their interaction with different learning spaces was, in part, influenced by both routine and time.

7.2   Routine and Time

In order to effectively implement SSP, the teachers were required to establish a variety of cooperative and collaborative relationships with other educators, their students and the wider school community. In other words, SSP required the teachers to establish a pedagogy, or a routine of practice, most conducive to providing socially-critical learning opportunities. For many of the teachers, this meant changing their previously well-established daily routine. The inability of teachers to achieve this change created rhetoric–reality gaps in the implementation of SSP.

Routines are unquestionably an essential part of daily life. As Giddens (1976) pointed out, routines, incorporating both established institutional processes and social customs and traditions, enable people to non-consciously act in ways that comply with social norms. Thus, each of the teacher's routine, at least in part, reflected their knowledge of the social norms associated with their work environment. The principals hoped that, by changing these social norms through the introduction of a new curriculum and new learning spaces, the teachers would be prompted, or motivated, to adjust their daily routines.

The belief that altering the teachers' routines was a potentially difficult task was held by the principals and teachers alike. Philip described "change" as "something that's very hard to do" due to well-established teaching routines: "some [of the teachers] are very regimented in the way they like going about things" and that as a result, "curriculum development hasn't changed in an eon". Fran suggested that well-established routines made changing pedagogy to be "especially challenging for teachers who have taught in the same classroom in the same way for twelve years or so", because routines act to maintain the status-quo. She agreed with the principals' assumptions that a significant change in the work environment might provide the much needed impetus for change, by motivating and thereby enabling the teachers to develop new routines or pedagogies. However, the development of the rhetoric–reality gaps in the implementation of SSP indicated that neither mandating a new curriculum (see Sect. 6.2.2), nor providing new learning spaces (see Sect. 7.1.1) motivated or enabled some of the teachers to alter their existing routines. Therefore, in order to better understand rhetoric–reality gaps in the implementation of SSP, it was important to investigate the pedagogies, as routines of practice, of the teachers. The teachers tended to describe their pedagogy as either a routine defined by a strict adherence to time, or a routine defined by a flexible approach to time.

The pedagogy that incorporated the strictest adherence to time was that practiced by Elizabeth. Time was central to Elizabeth's work, both in terms of her interpretation and implementation of SSP. She described the educational outcomes of SSP as "not a 9 to 4 concept" but "a 24 hour concept". This description was not inaccurate, but it nevertheless highlighted Elizabeth's propensity to establish meaning founded on the basis of time. Elizabeth's description of her efforts to implement SSP reflected a well-established and precisely timed schedule for waste management. Waste management incorporated a timetabled series of tasks to be completed by the students, based on the need to distribute and collect different types of bins from different areas of the school at specific times each week. Each task was timed, to ensure that it fitted precisely into Elizabeth's daily routine. She stated that "it's a huge task" to maintain such a routine, and that "it has to be well organised, otherwise it would really fall in a heap very quickly". Elizabeth's approach to SSP highlighted her preference for a well organised, and therefore predictable, work environment.

Irrespective of Elizabeth's pedagogical preferences, her classroom practices provided valuable insights into the potential of routines to influence educational rhetoric–reality gaps. Elizabeth initially developed the waste-management routine, in response to her previous principal's request, to enable Mountain Primary School to satisfy the SSP requirements for achieving five-star accreditation. Elizabeth stated that "I wouldn't have chosen to [do this as] it's a huge task" and described the organisational and time demands of the waste management routine as onerous: "logistically it's full on". These comments indicated that Elizabeth had developed her routine only because of the structural influence of the hierarchical management system of her work environment. Despite this, Elizabeth had chosen not to modify or abandon this routine even after the arrival of a new principal meant that SSP was no longer a school priority. In other words, there was a point in time at which Elizabeth considered it easier to maintain this difficult, but well-established, routine than to change it: the routine had become "institutionalized" (Fullan 2007, p. 65). Elizabeth's desire for a well-structured and predictable work environment supported by practiced daily routines outweighed her frustration or dislike of those same routines.

Elizabeth's case highlighted the effect of the strategy of establishing a new routine of practice in order to influence long term change. This strategy has been an important component of many social policy campaigns. Campaigns that attempt to provide information to encourage people to act in a particular manner are often not as successful as those which concentrate on getting the desired behaviours established, then explain why, as evidenced by the success of recent campaigns to reduce household water use in drought stricken Victoria (Kollmuss and Agyeman 2002). However, Elizabeth's experience of trying to alter behaviour led her to a different understanding of this strategy. Elizabeth candidly assessed the effect of her waste management routine as a strategy for behavioural change as poor. She recognised that simply telling people (in this case, the students) to follow a routine, especially one which had been enforced from a higher authority (a teacher), did not ensure behavioural change. She noted that her efforts to reduce rubbish and improve the management of waste within the school had not been as effective as she had expected, and was reticent to introduce new or improved rules or policies: "you wouldn't just introduce it because it wouldn't work". She believed that a higher level of compliance with the rubbish protocols within the school would require "more education…I think you really need to educate first". This suggested that Elizabeth considered things such as increased awareness as essential in establishing a new behavioural routine, and supported the notion that a change in teaching or learning "presupposes that both teachers and students share a common understanding of the new patterns of behaviour" (Gynnild 2002, p. 301). Similarly, Elizabeth recognised the role of motivation in changing behaviour, explaining that that was why she had introduced the "golden wheelie bin award…for the class that has got the lowest amount of waste" and that "each week every child with a waste free lunch gets a chance to win a prize".

The motivating factors (or possible sanctions; see Sect. 3.10), other than the principal's directions, that enabled Elizabeth to alter her previous routine in order to accommodate the waste-management schedule were not clear. It was evident however, that Elizabeth maintained a routine which was not only difficult and unpleasant to continue, but which also addressed a program no longer considered a priority by her school. This suggested that she did not enjoy change, and that, in line with Giddens' understanding of unconscious human motivation (see Sect. 3.9), she found ontological security through the maintenance of a well-organised and therefore predictable routine. Elizabeth's case demonstrated that for many people the reality of a well-established routine, even if it is less than ideal, is easier to maintain than to change. This highlighted the potential of routines to facilitate the development of educational rhetoric–reality gaps.

Similarly, the presence of a school bell strongly influenced the development of a strict time-directed daily routine for many of the teachers. For example, the shared 'recycle, re-use, reduce' lesson directed by Anita and Robyn at East Valley Primary School highlighted these teachers' desire to fit a particular set of learning activities into a time slot defined by the bell. Many of their students were obviously frustrated when they were not allowed to complete the tasks that had been set. Similarly, many of the students obviously rushed to complete a task rather than attempt to do their best work. In addition, Anita and Robyn completed certain aspects of tasks for the slower students in order to save time. It was not clear that the students successfully achieved the learning outcomes of the lesson identified by Anita and Robyn, because both the students and the teachers seemed unduly focused on time. In other words, Anita and Robyn not only directed the learning outcomes and learning activities for this lesson, but also the time in which it would take for the students to effectively master these outcomes. In addition, Anita noted that the classroom components of the implementation of SSP at East Valley Primary School had been timetabled to be undertaken at specific times: "the decision was made that we'd do sustainability in terms 3 and 4" as discrete biennial learning modules. This segregated SSP from the rest of the school curriculum, and effectively precluded the incorporation of sustainability ideals into the daily routine of the teachers and the students. This highlighted Giddens' notion of a duality between structure and agency, where the vocational/neo-classical approach to SSP was shaped, in part, by the timetabling of time-restricted learning activities which in turn, influenced the type of pedagogy most readily implemented (Giddens 1984).

The socially-critical pedagogy advocated by SSP was most readily implemented by the teachers, such as Karen and Cathy, who had a flexible approach to time as part of their usual routine of practice. This flexible approach also indicated that these teachers were more amenable to change. Unlike Elizabeth, Cathy and Karen both described their approach to SSP in terms of an open or negotiable timetable. Cathy for example, stated that she would happily abandon an entire learning program if the students were demonstrating enthusiasm for an alternative activity that offered equivalent learning opportunities. She noted that this approach ensured that "there's something new all the time, and I think that's what the beauty of it [SSP] is, things crop up all the time". In contrast to Elizabeth, Cathy indicated that such an approach was an essential contributor to her ontological security, stating that: "I couldn't do the same thing over and over and over and over again...I think I'd stagnate if I had to do the same thing over and over again". Not only did she indicate that a flexible routine "keeps life interesting" and "keeps me fresh", but that this was also essential for providing the best learning environment for her students:

> we [the teachers] have to be motivated to get the kids motivated, if we're not really excited about doing something, how can we make the kids excited about doing it, and I can't see that you [a teacher] can get excited about something that you've done twenty times before.

The most flexible attitude towards time, however, was demonstrated by Karen. At EVNP, Karen immersed herself and her students in the environmental realities of the out-of-school setting, stating that the "timetable is thrown in the wind, we don't have bells, we don't have loud speakers...I encourage children to work to their own time". Karen understood that the timetabled restrictions of teacher-directed learning was not an effective approach. Within the time that the students were present at the park, Karen provided support and guidance for the students to participate in the activities, or learning opportunities, in which they were most interested. As many of those opportunities arose from unexpected invitations or events within the park, they could not be predicted or timetabled. Similarly, Karen accepted that the learning from such opportunities could not be predicted or timetabled. Karen's ability to accept a flexible and dynamic timetable enabled her students to work collaboratively with each other and a range of people from the local community. The students' ability to take advantage of interesting and authentic learning opportunities as they arose ensured that they were learning within a socially-critical environment.

As discussed in relation to learning space (see Sect. 7.1), Karen reported that many of her colleagues who accompanied their students to EVNP found the learning space "extremely threatening", particularly due to the lack of facilities for organising time. Karen believed that most of those teachers sought a consistent and predictable work environment, and found the lack of school bells and the lack of times for specific forms of learning to be quite frightening. In other words, for some of the teachers, the physical aspects of a learning environment assisted them to undertake a routine dependent on organising time. Lisa was one of those teachers.

Lisa's perfectly organised classroom reflected her pre-planned pedagogy which, like Elizabeth's waste-management routine, was delivered in precisely timed portions. However, when asked what prevented her from implementing the socially-critical pedagogy that she recognised as essential for achieving SSP goals, Lisa stated "I think probably time". She explained that she found it difficult to organise her time because "there's always something going on" which causes many "general interruptions across the day". She explained that "trying to find those ways to get around it [interruptions]…can be a lot of organisation and management". Lisa believed that the only way in which to reduce the pressure of attempting to teach so many programs was to "integrate [learning outcomes] as much as possible". This answer was consistent with comments from all of the teachers, irrespective of their chosen pedagogy, that insufficient time, often due to an overcrowded curriculum and numerous special school programs, constrained their ability to improve or change their classroom practices. This was supported by the suggestion that "environmental education theory, as it is now, is not sufficiently grounded in teachers' experiences and in what they feel schools can do or what the school day is really like" (Robertson and Krugly-Smolska 1997, p. 323). Cathy for example, suggested that implementing SSP through a socially-critical pedagogy required her to establish and maintain collaborative relationships with people and organisations outside the school. This was not only the most difficult component of her work, but also required a significant investment in time: "time is definitely the killer—it really is". Similarly, Fran reported that her colleagues who were most resistant to introducing a socially-critical pedagogy actually feared the amount of time that they perceived such a change would require: "it's a fear rather probably than a resistance I think, a fear that they don't have time".

However, time is often a reason cited by teachers for not undertaking new practices (e.g. Barrett 2007; Palmer 1998; Tomlins and Froud 1994). The perfunctory manner in which time was identified as a problem by the teachers implied that such complaints were almost unconscious responses to an expectation, that is, a perceived social norm that teachers were busy people who were always stretched for time. David explained the reluctance of some teachers to participate in some programs as "teachers are all busy and there's always a pile of stuff we're not getting done". Lisa suggested that identifying time as a constraining factor was an "excuse" to explain ineffective or irrelevant aspects of a teaching routine, stating that change required "just re-organising the way things are structured or getting rid of things that aren't needed" and "leaving things behind that you don't need to be doing any more…that are blocking up the time, blocking up the space". Similarly, David noted that the choice to practice a socially-critical pedagogy could actually reduce the work load of a teacher by "empowering kids" with "authentic learning" experiences. He described a vocational/neo-classical approach as "too much work, we're busy enough as it is…we [the teachers] don't need this extra [planning] work when you've got kids who can do it…and parents and community". He noted that "allowing the kids to have some input" is not only "empowering to kids" and provides opportunities for "authentic learning", but it also reduces the planning or preparation work of a teacher by incorporating aspects of these into the learning process.

The teachers' routines of practice reflected different ways in which they related to time, and different ways in which they utilised learning spaces to implement a pedagogy that supported their relationship with time. For some of the teachers, in the context of implementing SSP, these relationships resulted in a rhetoric–reality gap. However, that is not to say that routines should not be part of a teacher's practice. Routines are an essential part of every teacher's practice. Classroom routines, for example, ensure that the students know how to handle normal daily occurrences: housekeeping routines enable the students to manage the physical components of a classroom, such as where to locate different learning materials; management routines assist the students to manage certain interactions, such as how to form a group; learning routines assist the students to approach learning in specific ways, such as reading quietly before writing an answer; and discourse routines provide rules for verbal exchange, such as raising a hand in order to ask the teacher a question and listening quietly while others talk (Leinhardt and Greeno 1986; Leinhardt et al. 1987). Such routines define the social norms of a classroom and ensure that the students understand a teacher's expectations (Burden 2003; Newsom 2001; Savage 1999). They therefore contribute to the students' feelings of ontological security, and reduce the need for teachers to micro manage every aspect of a classroom.

The difference between the use of a vocational/neo-classical pedagogy by teachers such as Lisa and Elizabeth, and the use of a socially-critical pedagogy by Cathy and Karen, was not the presence or absence of these types of routines, but the effect of routines on what might be considered "patterns of thinking", that is, the manner in which routines "support and scaffold" specific patterns of thinking (Ritchhart et al. 2006, p. 1). Both Cathy and Karen had taught their students to embrace patterns of thinking which incorporated the use of negotiation, cooperation and collaborative endeavour in order to identify: interests that may or may not be identical to those of their peers; ways in which to acquire information about those interests; and engaging ways in which to demonstrate their learning. In other words, many of the classroom routines established by Cathy and Karen were not a reflection of "ordinariness, habit and ritual" but "practices crafted to achieve specific ends" (Ritchhart et al. 2006, p. 5).

In the same way that the teachers used routines to establish the students' feelings of ontological security, it is easy to understand that routines were instrumental in assisting the teachers to establish ontological security for themselves. The educational rhetoric–reality gaps in the implementation of SSP undoubtedly reflected the relationship between a teacher's feeling of ontological security and the practice they were required to implement. The socially-critical pedagogy advocated by SSP was most successfully implemented by the teachers for whom a flexible approach to time was part of their usual routine of practice. The teachers who practiced routines heavily dependent on time not only found the socially-critical approach to SSP unfamiliar, but also seemed to consider the very notion of change to be challenging. Similarly, a socially-critical pedagogy was most successfully implemented by the teachers who designed routines that enabled the students to embrace negotiation, collaboration and cooperative learning as part of their normal learning routine. The teachers who taught to routines heavily dependent on the continuous provision of directions to their students could not implement a socially-critical approach. The practices of those teachers were most likely to represent a rhetoric–reality gap.

However, that is not to say that those teachers were incapable of change. As indicated by the development of the waste-management routine by Elizabeth, appropriate motivation (or sanction) could enable teachers to alter (or maintain) well-established routines.

Although this discussion has focused on the need for the teachers to alter their pedagogical routines, it is important to note that the implementation of SSP also required the teachers to alter routines related to the subjects, or content, that they routinely taught. David acknowledged that these routines were particularly influential in some teachers' ability, or willingness, to embrace SSP at Ocean Primary School:

> it's a common understanding [that every primary school teacher teaches maths]…there'd be no one here who wouldn't teach maths…some teachers might teach it less, or less enthusiastically…but no one would think of not doing it…and if they wanted help they'd get it… they'd maybe collaborate with other teachers and they'd maybe use those worksheets so that they can have a cheat sheet and cover their misunderstanding or not understanding.

In other words, David acknowledged that many of the teachers at Ocean Primary School felt unable to incorporate SSP into their teaching routine, not just because of the requirement to enact a socially-critical pedagogy, but also because the ideals and content of environmental, or sustainability, education were not part of their usual teaching routine.

Although the well-established routines employed by the teachers undoubtedly contributed to the development of the rhetoric–reality gaps during the implementation of SSP, they did not fully explain such gaps. In order to better understand such rhetoric–reality gaps it was essential to understand the other ontological elements that significantly constrained the teachers' ability to embrace change.

7.3   Other Resources

In order to effectively implement SSP through a socially-critical pedagogy, the teachers were required to: re-define their roles in the learning process; re-define what constituted a learning space; establish a variety of cooperative and collaborative relationships with other educators, students and the wider school community; and in general, establish a routine of practice most conducive to providing socially-critical learning opportunities. The presence of rhetoric–reality gaps in the implementation of SSP indicated that many of the teachers were unable to do these. Most of these teachers suggested that their inability to implement a socially-critical pedagogy was due, in part, to the lack of certain resources—a reason often offered by the teachers to justify the lack of environmental education in schools (e.g. Barrett 2007; Palmer 1998; Tomlins and Froud 1994). The role of resources in the development of rhetoric-reality gaps is discussed in terms of: allocative resources, or physical teaching and learning aids such as science equipment (Sect. 7.3.1); and authoritative resources, such as the expertise of others (Sect. 7.3.2). Several of the teachers attributed the lack of these types of resources to insufficient funding.

7.3.1 Allocative Resources

According to Giddens, an unequal distribution of allocative resources, such as equipment used for certain teaching and learning activities, can contribute to unequal human relationships, which in turn can influence a teacher's capacity to act in a particular manner (Giddens 1979; Turner 2003). Several of the teachers commented on their perception of the inequality of the state (Victorian Government) education system in terms of allocative resources. Andrew, for example, lamented that "resources are our biggest issue out here. Other schools have things like microscopes…we don't have the opportunity to use those kinds of things". Other teachers considered that the lack of allocative resources contributed to them having insufficient time to plan and organise more effective teaching practices. Simon for example, noted that he would benefit from access to some "ultra organised cupboards with lots of stuff in them" stating that "a lot of my time in science is spent getting stuff together". However, the lack of these types of allocative resources alone did not adequately explain the rhetoric–reality gaps that developed during the implementation of SSP. Even Elizabeth considered such resources unrelated to the implementation of a socially-critical pedagogy, stating that the potential for the lack of these resources to inhibit a teacher's practice: "would depend on what your goals were for teaching…if it was sharing of information and sharing of learning, and designing student-centred classroom tasks, no it wouldn't inhibit it at all". Similarly, David circumvented a lack of physical resources for certain projects through the implementation of a socially-critical approach which encouraged the students to find ways of making their own equipment, negotiating for assistance, or raising funds to purchase necessary materials for SSP-related projects. He believed that this was a valuable approach which helped the students to develop a critical awareness of the real world, stating that "we don't want the kids to think everything's laid on for them…they've got to run what's going on out there". Elizabeth summarised this ideal: "the whole idea about sustainability in environmental education is that you re-use and use, and use well the resources you've got, not go out and pluck new resources". In other words, the degree to which any teacher had access to specific allocative resources neither enabled, nor constrained, their ability to implement SSP through a socially-critical pedagogy. Most of the teachers however, suggested that the most critical resources for implementing SSP were not physical resources, but included the knowledge and skills, or expertise, of others. These were authoritative resources.

7.3.2 Authoritative Resources

According to Giddens, non-physical, or authoritative resources, relate to an individual's capacity to influence, direct or organise various aspects of social interaction, such as time and space (as discussed earlier; see Sects. 7.1 and 7.2) or association (Giddens 1979; Turner 2003). The notion that people were valuable resources, and that collaborative teaching and learning provided access to, for example, the expertise of others, was central to effectively implementing SSP through a socially-critical pedagogy. Such expertise was sought to assist the teachers to improve their pedagogy and assist the students to improve their learning. The former related to perceived level of teacher support, or association, as discussed in Sect. 6.5.1 (Taylor 2003; Arts 2000), while the latter, discussed here, related to resources that the students could access. In the context of SSP, the students accessed such resources only when the teachers assisted them to participate in collaborative, community-based and multi-age learning experiences. However, despite the fact that the teachers indicated a good understanding of the ability of a socially-critical pedagogy to provide such resources for their students, few fully embraced such an approach.

The teachers who most effectively embraced a collaborative teaching and learning environment, Karen and Cathy for example, did not believe that their students required additional allocative or authoritative resources. Karen's socially-critical approach to SSP centred on collaborative efforts between the students and the staff at EVNP. The students undertook a wide range of caretaker and scientific roles through working cooperatively and collaboratively with EVNP personnel, members of the public and various government agencies. As these students were participating in real world activities, the experts with whom they worked provided not just equipment, but also specialised knowledge, ideas and opinions. Karen used simple learning activities that were not dependent on physical resources to assist the students to maintain these relationships, including the establishment of a postal network between the students and the EVNP personnel. This network facilitated an ongoing exchange of ideas and information between visits to EVNP. Cathy also assisted the students to establish a range of collaborative learning relationships. In order to effectively develop a student-initiated bird breeding program, for example, Cathy assisted the students to seek help from various educators and scientists with specialist biological knowledge and nest box building expertise. Although the students applied for a grant in order to purchase materials for making nesting boxes, the project could not have progressed without the sharing of knowledge between the students and several bird experts. It was evident to both Karen and Cathy that the learning opportunities provided by these types of collaborative experiences far outweighed the benefits that just additional physical resources could achieve.

Some of the teachers used guest educators as an initial step in moving away from a vocational/neo-classical pedagogy. Andrew and Lisa for example, asked field educators associated with a local water authority to direct certain science lessons, stating that: "we try and use these as much as possible, as much as we can, [be]cause obviously they know more about water than we do" (Andrew). Similarly, Julia sought assistance from the science teaching staff at a neighbouring secondary school to enable her students to experience aspects of science she was not confident to teach. These experiences represented a significant change for the teachers who, as eloquently expressed by Philip, previously believed "I am the font of all knowledge and I spew forth".

However, other teachers identified the lack of student resources as a contributing factor in their inability to implement SSP. Andrew, for example, explained that "we have small classes...to book a bus to go somewhere...it's a high expense to the kids". He stated that, with additional money "the resources that we could then use... take kids here, take kids there". Although Andrew noted that "it's terrific that we are in a rural situation...we do have the river to go and visit...a river at the back door" he sought money in order to transport students to other locations with resources such as "bird habitats", or facilities for "water-based activities [such as] water testing and pond life". Andrew did not view a socially-critical pedagogy as a way in which to engage students with learning within the school and surrounding environments. This not only suggested that Andrew held specific ideas about what constituted an appropriate space for learning (see Sect. 7.1) but that he also used the lack of money to justify his inability to fully implement SSP.

Similarly, Elizabeth stated that the lack of funds at Mountain Primary School meant that she was unable to provide opportunities for the students to participate in certain activities such as "research in a true scientific sort of way, or...hands-on activities that involved excursions, or paid guest speakers". This comment not only suggested that Elizabeth did not understand that science was first and foremost "a process of generating information" (Foulds and Rowe 1996, p. 16), but also provided valuable insights into the teachers' complaints regarding their inability to expose the students to the expertise of others. Elizabeth had almost unlimited access to the school kitchen garden managed by an expert horticultural manager, and nearby state parks with dedicated education officers. Her belief that opportunities for the students to learn from others required "paid guest speakers" was shared by other teachers, including for example David, who wanted additional funds in order "to buy in people". These comments suggested an unwillingness to assist the students to negotiate and collaborate with others in order to establish relationships, from which learning from others would occur naturally. In other words, these teachers viewed funding as a means through which to provide resources for the students, in terms of access to people, without having to significantly adjust their usual pedagogical routines. Money was viewed by some of the teachers as a resource that provided them with the power to avoid change.

It is important to note that several other teachers indicated that the lack of resources, in terms of people who "are expert at things" (Lisa), contributed to their difficulty in implementing SSP, but did not relate this to financial shortfalls. Lisa, for example, explained that South Bay Primary School had been attempting to "draw more parents in for different roles" and "call on different people...to do different things". Although these efforts represented attempts to increase the level of community involvement in the school, Lisa's comments indicated that this fell short of offering the collaborative learning opportunities for students that SSP intended. She explained that the school was actively "encouraging other people to feel welcome to come into the school" because "a lot of new parents that come in feel intimidated or pressured not to be a part of it [the parent body]...it's the same ten parents that do it...a small community [with] quite clicky groups". Lisa also explained that certain parents within this group had shouldered the responsibility for the design and construction of outdoor learning areas for the implementation of SSP, including the "veggie garden" and the "frog pond", and that now it was time for "different parents" to contribute. In other words, the school viewed the community as a resource for the development of outdoor learning spaces *for* the students, rather than *with* the students.

Thus, despite the perceived disadvantages of an apparent lack of resources identified by some of the teachers, access to additional resources was not essential for implementing SSP. The teachers who most effectively implemented SSP, like Karen and Cathy, embraced a socially-critical pedagogy as a way in which to establish cooperative and collaborative relationships which provided opportunities for the students to learn through participation, that is, through the sharing of ideas and knowledge. These teachers did not rely on purchasing power to acquire people as resources, but assisted the students to explore different ways in which to access the people, or knowledge, or skills, most suited to their interests and chosen projects. In contrast, the teachers who tried to implement SSP through a vocational/neo-classical pedagogy, like Elizabeth and Lisa, failed to assist or encourage the students to access the expertise of any other people, either from within the school or the local community.

The teachers' responses to the requirement to implement SSP provided valuable insights into the complex relationship between authoritative and allocative resources, and how a teacher's perception of the resources available to them will influence their students' learning experiences. Implementation of a socially-critical pedagogy, by a teacher, meant that students gained opportunities to access a variety of both allocative and authoritative resources, which increased their confidence in building relationships, and therefore assisted them to create further opportunities to access additional resources. This highlighted Giddens' notion of a duality between structure and agency, where the socially-critical pedagogy experienced by the students, was shaped by the resources accessed, and in turn, influenced the types of resources sought (Giddens 1984). In all cases, the teachers held the authority to give their students access to resources through implementing SSP, indicating that access to resources for the students neither constrained, nor enabled, the implementation of a socially-critical pedagogy by the teachers. In other words, the rhetoric–reality gaps that developed during the implementation of SSP did not simply reflect the teachers' inability to access appropriate resources for their students.

7.4   Duality of Structure and Agency and Educational Rhetoric–Reality Gaps

Analysis of the rhetoric and the reality of the teachers who were required to implement SSP and a socially-critical pedagogy demonstrated the effect of the duality of structure and agency (Giddens 1984) on those teachers' practices, and highlighted some of the causes of the educational rhetoric–reality gaps that developed as a result of the implementation of this program.

The teachers understood both the environmental and educational goals of SSP. The teachers' ideas regarding the potential for SSP to influence their own lives as well as the lives of their students and the broader school community demonstrated their understanding of the future-oriented and socially-transformative goals of SSP, and that it addressed purposes of education best described as democratic equality (Labaree 1997). The principals shared these understandings, but indicated that their decision to implement SSP was also based on its potential to operate as a vehicle for pedagogical change. This highlighted the way in which different structural elements, in this instance a 'structured set', could represent different things to different people. In this instance, the principals used their hierarchical position to define certain aspects, or rules, of the environment in which the teachers worked. Irrespective of directions given by the principals and the rhetoric provided by SSP documents, the teachers' practices indicated that they approached the implementation of SSP in one of two ways: (i) the teachers modified and adjusted the structural components of their working environment in order to enable them to engage their students through a socially-critical pedagogy, or; (ii) the teachers modified and adjusted the implementation of SSP to suit the existing structural components of their working environment.

Cathy, for example, did not permit the physical conditions of her work environment to constrain her use of a socially-critical pedagogy. She encouraged the students to determine how to best utilise existing resources, and to identify and use new and different learning spaces when appropriate. Cathy also adopted a flexible approach to other aspects of her working environment, such as time. She indicated that she would only allow a specific curriculum to influence her teaching if the students were engaged and learning, and would happily extend or forego planned curriculum-based activities in response to the students' learning needs and interests. Similarly, Karen encouraged the students to take responsibility for their time at EVNP, not just in terms of planning their usual daily activities, but most importantly, in relation to identifying and creating opportunities to participate in, and learn from, real life experiences. In other words, both Cathy and Karen established a routine which embraced flexibility, openness to the students' needs and interests, and a willingness to engage with the learning opportunities provided by real life experiences as they arose. Such routines were not defined by structured sets, rules or physical resources. Such routines established a social norm in which the students attended school with the expectation that they were responsible for learning in an environment which incorporated a certain level of negotiation, collaboration and cooperation. These routines embraced the notion that new interests and opportunities, or changes, were an integral component of life and learning and school.

In contrast, teachers such as Lisa and Elizabeth permitted various structural elements of their work environment, particularly the physical aspects of their classroom learning spaces and the use of time, to define their pedagogy. Lisa and Elizabeth established routines in which curriculum-directed learning occurred through planned activities undertaken in set ways within certain learning spaces at specific times. Such routines established a social norm in which the students attended school with the expectation that their teachers had determined what they would learn, how they would learn it, how long they needed to learn it, and where that learning would take place. The ability to maintain such a routine demanded that any additional or different activities were planned and completed within an allocated time. By definition, the social norm established by these routines did not encourage or embrace change, because even a small change had the potential to impact not only on the plans for a single day, but also for an entire school term. Both Lisa and Elizabeth attempted to implement SSP through their existing routines.

In other words, the implementation of SSP demonstrated that, once established, a teacher's routine of practice effectively operated as a self-supporting, or self-fulfilling, system. Each routine defined the manner in which the teachers and the students interacted with each other and the world while at school. Each routine defined the social norms for learning and teaching at school, which, when practiced, defined that routine. This is the essence of Giddens' notion of the duality of structure and agency (1984).

Although the rhetoric–reality gaps in the implementation of SSP were formed by the practice of routines which demonstrated the way in which structure and agency operated as a duality, that duality of structure and agency did not cause these rhetoric–reality gaps. Analysis of the rhetoric and reality of the implementation of SSP by the teachers showed that neither the presence, nor the absence, of ontological elements such as new and different learning spaces, physical resources, perceived principal and peer support, or even time, predicted whether or not the teachers implemented SSP through a socially-critical pedagogy: the structural features of the school work environment did not universally constrain, or enable, the teachers to implement a socially-critical pedagogy. However, the teachers' stories indicated that their beliefs about the environment, and beliefs about education influenced their perception of SSP goals, whether or not they embraced SSP principles in their own lives, and the manner in which they chose to implement SSP in their classrooms. Thus, the development of the educational rhetoric–reality gaps, in the context of the implementation of SSP, was an issue of teacher agency.

Thus, in order to identify a possible intervention point, or ontological element, through which activities and/or policies designed to reduce the development of educational rhetoric–reality gaps could be introduced into an institutional environment in which teachers work, it was essential to identify the critical aspects of agency that influenced the teachers' pedagogical decisions. Analysis of the teachers' agency, most particularly in terms of the teachers' environmental ideology and educational ideology, and the relationship between ideology and structuration ontological elements is discussed in Chap. 8.

References

Arts, B. (2000). Regimes, non-state actors and the state system: A 'structurational' regime model. *European Journal of International Relations, 6*(4), 513–542.

Barrett, M. J. (2007). Homework and fieldwork: Investigations into the rhetoric-reality gap in environmental education research and pedagogy. *Environmental Education Research, 13*(2), 209–223.

Burden, P. R. (2003). *Classroom management: Creating a successful learning community.* New York: Wiley.

Foulds, W., & Rowe, J. (1996). The enhancement of science process skills in primary teacher education students. *Australian Journal of Teacher Education, 21*(1), 16–21.

Fullan, M. (2007). *The new meaning of educational change* (4th ed.). New York: Teachers College Press.

Giddens, A. (1976). *New rules of sociological method.* New York: Hutchinson.

Giddens, A. (1979). *Central problems in social theory: Action, structure and contradiction in social analysis.* London: Macmillan.

Giddens, A. (1984). *The constitution of society.* Cambridge: Polity Press.

Gynnild, V. (2002). Agency and structure in engineering education: Perspectives on educational change in light of Anthony Giddens' structuration theory. *European Journal of Engineering Education, 27*(3), 297–303.

Kollmuss, A., & Agyeman, J. (2002). Mind the gap: Why do people act environmentally and what are the barriers to pro-environmental behaviour? *Environmental Education Research, 8*(3), 239–260.

Labaree, D. F. (1997). Public goods, private goods: The American struggle over educational goals. *American Educational Research Journal, 34*(1), 39–81.

Leinhardt, G., & Greeno, J. (1986). The cognitive skill of teaching. *Journal of Educational Psychology, 78*(2), 75–95.

Leinhardt, G., Weidman, C., & Hammond, K. M. (1987). Introduction and integration of classroom routines by expert teachers. *Curriculum Inquiry, 17*(2), 135–175.

Lippman, P. C. (2002). Just a thought. Practice theory, pedagogy, and the design of learning environments. *The Quarterly Newsletter of the Committee on Architecture for Education, 2*(July 2002), 1–7.

Newsom, J. (2001). Sanctuary or sanction? *American School Board Journal, 188*(7), 24–28.

Palmer, J. A. (1998). *Environmental education in the 21st century: Theory, practice, progress and promise.* New York: Routledge.

Ritchhart, R., Palmer, P., Church, M., & Tishman, S. (2006, April 7–11). *Thinking routines: Establishing patterns of thinking in the classroom.* Presented at the annual meeting of the American Educational Research Association, San Francisco, USA. http://www.ronritchhart.com/Papers_files/AERA06ThinkingRoutinesV3.pdf. Accessed 8 Oct 2014.

Robertson, C. L., & Krugly-Smolska, E. (1997). Gaps between advocated practices and teaching realities in environmental education. *Environmental Education Research, 3*(3), 311–326.

Rudd, T., Gifford, C., Morrison, J., & Facer, K. (2006). *What if…? Re-imagining learning spaces* (Opening education). Bristol: Futurelab.

Savage, T. (1999). *Teaching self-control through management and discipline.* Boston: Allyn and Bacon.

Taylor, V. J. (2003). Structuration revisited: A test case for an industrial archaeology methodology for far North Queensland. *Industrial Archaeology Review, XXV*(2), 129–145.

Tomlins, B., & Froud, K. (1994). *Environmental education: Teaching approaches and students' attitudes: A briefing paper.* Slough: UK Foundation for Educational Research.

Trautmann, N. M., & MacKinster, J. G. (2005, January 19–23). *Teacher/scientist partnerships as professional development: Understanding how collaboration can lead to inquiry.* Paper presented at the International Conference of the Association for the Education of Teachers of Science, Colorado Springs, USA. http://ei.cornell.edu/pubs/AETS_CSIP_%2005.pdf. Accessed 1 Oct 2014.

Turner, J. H. (2003). Structuration theory: Anthony Giddens. In *The structure of sociological theory* (7th ed., pp. 476–490). Belmont: Wadsworth.

VanNoteChism, N. (2006). Challenging traditional assumptions and rethinking learning spaces. In D. G. Oblinger (Ed.), *Learning spaces.* Washington, DC: EDUCAUSE.

Chapter 8
Ideology and Ontological Security

The implementation of the Sustainable Schools Program was accompanied by the development of educational rhetoric–reality gaps, influenced most significantly by the teachers' perceptions and experiences of the practicalities of implementing a socially-critical pedagogy. The major differences between the teachers whose classroom practices defined best practice and those whose practices represented a rhetoric–reality gap were best described as aspects of teacher agency. This chapter identifies important relationships between the beliefs held by the teachers, the values embedded within the goals of the Sustainable Schools Program and the practice of a socially-critical pedagogy, and the pedagogical practices that the teachers chose to employ when asked to implement the program. This discussion draws on Giddens' theory of structuration to highlight the critical elements of such relationships, particularly in terms of the teachers' environmental and educational ideologies and the notion of ontological security, and the role of these elements in the development of the educational rhetoric–reality gaps that accompanied the implementation of the Sustainable Schools Program—here identified as 'ideological rhetoric–reality gaps'. Understanding the manner in which 'ideological rhetoric–reality gaps' contribute to the duality of structure and agency in a teacher's classroom practice assists to identify potential intervention points for reducing the prevalence and/or severity of such educational rhetoric–reality gaps in the future.

© Springer International Publishing Switzerland 2016
J. Edwards, *Socially-critical Environmental Education in Primary Classrooms*,
International Explorations in Outdoor and Environmental Education 1,
DOI 10.1007/978-3-319-02147-8_8

8.1 Ideology

The term 'ideology',[1] as used here, refers to the beliefs about the way in which a society "ought to function to support the livelihoods and/or aspirations of its members" (Short and Burke 1996, p. 14). Manno (2004) noted that the dominant ideologies of modern western societies "have been those that prescribe the role of the individual, the community, and the state in relation to the society-shaping forces inherent in capitalism" (p. 158). Thus, in the context of this discussion, the environmental goals of the Sustainable Schools Program (SSP) and the understanding of issues arising from human–environment relationships, were most likely to be perceived through the lens of the well-established social norms of Australian society (Sunderlin 2003). Similarly, the educational goals of SSP, particularly in relation to the need to implement a socially-critical pedagogy, were most likely to be perceived through the lens of the well-established social norms of the teachers' work environments. The rhetoric of the teachers who were required to implement SSP indicated that differences in the ways in which they understood such issues reflected the degree to which they assigned intrinsic value to the natural environment (environmental ideology; a narrow view of Education *for* Sustainable Development, ESD, but one which reflects the SSP focus on environmental education, as outlined in Sect. 2.4), and viewed their role as a teacher (educational ideology). This was supported by the notion that beliefs, attitudes and values "underlie the stances teachers…adopt when analysing realities, challenging constraints, and promoting excellence for all" (Butcher and McDonald 2007, p. 12). Similarly, the role of the teachers' personal ideologies in the implementation of new curricula was highlighted by an investigation into reports of stress amongst teachers in the United Kingdom, which found that "individuals' attitudinal responses to change are determined by the extent of compatibility between their own ideologies, values and beliefs and those reflected in the changes they encountered" (Evans 2000, p. 185). Investigating the environmental and educational ideologies of the teachers implementing SSP was therefore an important step in identifying factors that could influence the development of rhetoric–reality gaps.

8.2 Environmental Ideology

Prior to implementing SSP, all of the teachers attended professional development sessions during which they explored the underlying environmental and educational values of the program. This was considered an important process, because implementing any "curriculum involves putting into action a system of beliefs. Therefore,

---

[1] The use of the term 'ideology' has a complex history which encompasses myriad definitions, a review of which is beyond the scope of this discussion. The definition adopted here was chosen for its ability to acknowledge and encompass the widest range of factors that influence and motivate educational endeavours in Australia.

when we engage in inquiry about curriculum, we examine our beliefs as well as our actions in the classroom" (Short and Burke 1996, p. 97). The rhetoric of the principals and teachers revealed the beliefs, and in particular, the environmental values (e.g. Sects. 6.2.1 and 6.3.1) that they perceived to be embedded in the goals of SSP (refer to Sect. 2.4 for discussion regarding the relationship between ESD and SSP).

Cotton (2006) proposed that "it is possible that teachers' pedagogical and environmental beliefs are more important in guiding their teaching about controversial environmental issues than have previously been recognised" (p. 69). Although the role of teachers' environmental ideologies in determining classroom practices is not well understood, it is true that "many environmental issues are controversial, at least in part, because of the differing attitudes and values held by interest groups" (Cotton 2006, p. 70). Thus, as the principals and the teachers demonstrated a good understanding of the goals and environmental values of SSP, it was important to investigate the relationship between their pedagogical choices for implementing the program and their personal environmental ideologies, in order to determine if these relationships contributed to the development of the rhetoric–reality gaps in the implementation of SSP.

A detailed discussion of environmental ideology, a subject of continuing and vigorous debate (e.g. Keller 2010; Rai et al. 2010; Vincent 2010) is beyond the scope of this account. However, as each teacher's practice reflected, in part, a relationship between their personal environmental ideology and that which they attributed to the goals of SSP, it is necessary to define the way in which the term environmental ideology is used here.

An individual's environmental ideology reflects their beliefs, attitudes and values, and falls within one of two broad perspectives: an anthropocentric perspective, in which nature has only extrinsic instrumental value as a resource (e.g. economic, recreational, scientific, historical or religious) for the benefit of humanity; or an ecocentric perspective, in which nature has intrinsic value (value in and of itself) unrelated to its perceived potential to contribute to human welfare (Vilkka 1997). Anthropocentric perspectives are particularly overt in capitalist industrialised cultures where scientific and technological developments exploit natural environments for personal and commercial human interests. This is reflected by the rhetoric of sustainable development, described by Vilkka (1997) as being representative of an "ideology of strong anthropocentrism which ignores our dependency on nature by mastering and dominating it" (p. 71).

As outlined in Chap. 2, SSP documents framed environmental education as future-oriented, socially-transformative activity that aimed to address certain aspects of current human–environment relationships. The teachers and the principals believed that SSP had been developed in response to concerns that humans were facing significant lifestyle and survival problems due to environmental damage caused by the unmitigated and unsustainable use of natural resources. They believed that SSP framed these concerns as activities that were both suitable and accessible to students of primary schools, and that this provided the potential for SSP to influence predominant human–environment relationships (see Sects. 6.2.1 and 6.3.1). Teachers identified the socially-transformative goals of SSP as beneficial for both current and future human lives, as stated by Simon for example, for the "the needs of people". All of the teachers identified anthropocentric values embedded in the goals of SSP. However, the teachers who practiced vocational/neo-classical and liberal-progressive pedagogies referred to the environment in a different manner to those who implemented a socially-critical pedagogy. This suggested that the implementation of SSP may have been influenced by the degree to which the environmental ideology held by each of the teachers correlated with that which they perceived to be embedded within SSP. These ideas are discussed in relation to the use of the terms *the* environment and *our* environment.

8.2.1   The *Environment*

Teachers who implemented SSP through vocational/neo-classical and/or liberal-progressive pedagogies, and who supported the use of a knowledge-based science curriculum for environmental education, consistently objectified the environment and held environmental ideologies that most strongly embraced anthropocentric perspectives. Andrew, for example, attributed the need to establish a sustainable future in order to ensure the protection of human life. Fran agreed, stating that she hoped to encourage students to make "changes that are…favourable to the environment", but that such changes shouldn't "mean that our lives are worse off". She noted that although humans have "a responsibility to change the way we're currently operating for environmental reasons…we are part of the environment…so we want it in-check for ourselves…it's for selfish reasons as well". Fran compared the understandings developed by the students undertaking SSP to "having the right to looking after themselves and then having the right to a better world". In other words, these teachers positively justified the goals of SSP as education *for* social transformation that would benefit humanity. They acknowledged the responsibility of humans in human–environment relationships, and recognised that their personal values, or environmental ideologies, could be considered in ways that were consistent with SSP goals.

However, one teacher in particular had significant difficulty aligning his environmental ideology with the goals of SSP. Simon believed that human needs must be the central focus of any education. In light of this, he was very concerned about participating in SSP due to:

> the negatives that come with the term environmentalist…I'm not the guy…who's going to rally if they don't shut the Gordon River Dam…I don't care if they cut some trees down, yeah, they're widening the road, I understand that, I'm not [going to] freak out [be]cause they're cutting trees down…the needs of people I think have to be weighed very carefully against [environmental needs].

Simon was most eager to qualify his stance on environmental issues, particularly in relation to environmental activism, and his belief that human needs must always take priority. As SSP professional development sessions were always carefully managed to avoid creating the feeling that teachers or students were required to adopt such activist roles, Simon's comments most likely reflected deeply held beliefs, attitudes and values. Several researchers (e.g. Greenall Gough and Robottom 1993; Robertson and Krugly-Smolska 1997; Simmons 1991) have suggested that teachers who consider environmental issues to be somewhat controversial may shy away from participatory learning activities that form part of any education *for* the environment. It is important to note that in this context, Simon's comments were directed towards environmental activism, not a socially-critical pedagogy, or "transformative teaching as a form of activism in teaching" (Matthews 2005, p. 95). Despite his strong feelings against environmental education, he did state that, having completed the SSP professional development sessions, "I'm not really slanted in that direction [environmental education], but I am interested in that now…from a really balanced…kind of angle, which I think is important". This suggested that Simon interpreted SSP as a framework that enabled him to, at least loosely, justify his participation in environmental education by balancing his anthropocentric values with the notion that environmental education could in some way benefit human lives. In other words, the SSP professional development sessions may have assisted Simon to find a way in which to relate his values to those embedded in the SSP goals, as "changing teachers' perceptions and understandings of the subject being taught may well change the values they can emphasise in class" (Bishop et al. 2005, p. 158). However, Simon's comment also suggests that, despite his anti-environmental views, he believed that he could implement SSP by maintaining a neutral, or 'balanced', position.

In a study of the way in which teachers incorporated controversial environmental issues into the teaching of geography, Cotton (2006) concluded that "teachers' beliefs are at odds with much published discourse on environmental education" because although the curriculum being followed advocated the "promotion of positive attitudes towards the environment, this agenda is not shared by teachers…[who] aimed at offering a 'balanced' picture of controversial environmental issues" (p. 77). This suggested that teachers, like Simon, might have dealt with curriculum and educational goals which contradicted their personal ideologies by deliberately aiming to exclude specific values from their classroom practices. However, as noted by Cotton (2006), "teachers' beliefs about balance" are not only "highly complex" but also "problematic in terms of their potential for translation into practice in the classroom" (pp. 72–73). As discussed in Sect. 2.3.4, the notion that "no education is politically neutral" (Wink 2000, p. 77) has been well-established in the literature (e.g. Fien 1999; Schugurensky 2002; Swain 2005; Wink 2000), and that as such, the "idea of maintaining a neutral position [in the classroom is] an illusion" (Cotton 2006, pp. 72–73).

Simon's case highlighted the fact that teachers are often expected to work towards goals that contradict their personal values. Simon explained that his values were not considered when his principal gave him the role of SSP coordinator at West Quay Primary School:

the principal chose a group of teachers to get involved with it [SSP]…I was chosen and Fran [a colleague] was chosen, probably because of our roles…and our grade levels. My role is science and that fits in quite well with it…hers is sort of an age group that this sort of works well with…grades 4 and 5…and I wouldn't say that it would have got very far without the school making it a definite goal.

Simon's comments provided valuable insights into the role of human values in educational change, particularly the relationship between the values held by the school principals and the teachers. As noted in Sect. 6.2.1, the principals were responsible for the decision to implement SSP. Simon's comments indicated that, because the goals of SSP did not reflect environmental ideologies held by the teachers at West Quay Primary School, it was being implemented only because the teachers were following their principal's instructions. These instructions, in turn, required the teachers to design learning activities for goals that contradicted their personal values. Simon, for example, clearly perceived the form of environmental education encouraged by SSP to represent values held by those engaged in radical environmental activism, and which focused on the need to preserve the natural environment at all costs. Simon perceived this environmental education to place the needs of the environment before the needs of humans, and this contradicted his personal anthropocentric values. He found it difficult to justify his participation in the implementation of SSP, and this caused him to feel extremely uncomfortable in his role as the coordinator of the program.

When there is such a significant disparity between a teacher's values and the values they perceive to be embedded in the curriculum they are implementing, educational change may be unsustainable. This was highlighted by Elizabeth and the implementation of SSP at Mountain Primary School. As noted in Sect. 5.3.1, despite the fact that Mountain Primary School had held SSP five-star accreditation for 2 years neither the principal (undertaking her second year at Mountain Primary School) nor the majority of the staff were aware of the program or its current status within the school. Elizabeth reported that SSP had been introduced into the school by the previous principal who had voiced a personal interest in environmental issues. When the current principal was asked to discuss the role of SSP in the school, she responded, "what is this Sustainable Schools Program?" and "I don't know if we do this thing here".

Elizabeth admitted that her involvement with SSP was neither voluntary nor enjoyable, and that she "wouldn't have chosen to" participate if she had been given the choice. She also indicated that other staff at Mountain Primary School had stopped implementing activities they had set up as part of SSP, such as the "energy-wise education program". The reasons for the inability (or unwillingness) of the teachers to continue the educational changes established as part of SSP under the auspices of the previous principal were not fully investigated. However, informal discussions with several of the teachers at Mountain Primary School, indicated that, like Simon, they perceived the values embedded in the goals of SSP to contradict both their own environmental ideologies, and their understanding of what constituted primary school education. Both Elizabeth and her principal were somewhat dismissive of the concept of environmental education. The notion that teachers may not consider environmental education to be a valid educational endeavour has been noted elsewhere. For example, when investigating ways in which to develop pre-service teachers' skills in environmental education, Cutter (1998) found that an overwhelming proportion of the pre-service teachers did not actually consider environmental education to be an important component of school education. In other words, the teachers at Mountain Primary School, like those at West Quay Primary School (see Simon's comments above) had implemented SSP only to fulfil a principal's expectations. Although those expectations reflected that principal's personal values, they did not necessarily reflect the values held by the teachers, and as a result, the teachers reverted to their original practices once those expectations were removed.

In addition, Simon's statement that "my role is science and that fits in quite well with it [SSP]" highlighted another way in which values may have influenced the teachers' practices, and provided an example of the manner in which Giddens' (1984) notion of the duality of structure and agency related to the development of educational rhetoric–reality gaps. Simon's statement reflected the idea that "teachers' values in the classroom are shaped to some extent by the values embedded in each subject, as perceived by them" (Bishop et al. 2005, p. 158). Although research regarding the role of teachers' values in the teaching of environmental education is limited, the role of teachers' values in the teaching of mathematics and science is a growing field of research (e.g. Bishop et al. 2005; Bishop et al. 2006; Clarkson et al. 2005). Such research has focused on the notion that teachers' personal values not only influence their views about a subject, but also their pedagogical choices for that subject. In turn, the values embedded in the pedagogical approaches chosen by teachers portray certain sets of values about a subject to their students (Bishop 2008; Chin et al. 2001). Simon held strong anthropocentric environmental values and consistently objectified the environment. Not only did those values influence Simon's perceptions of the values embedded in the goals of SSP, but also led him to consider environmental education to be part of a traditional science education, that is, education *about* the environment (Lucas1980; see Sect. 2.2.1).

As discussed in Sect. 2.3.2, although science knowledge and environmental education should not be considered mutually exclusive (Gough 2007), it is true that traditional science pedagogy, represented here by a vocational/neo-classical approach, does not fully support the future-oriented and socially transformative outcomes of SSP. The use of a vocational/neo-classical pedagogy not only objectifies the natural environment, but also separates humans from their environment and segregates facts from values (Scott and Gough 2004). The teachers who consistently objectified the environment were those who also employed a vocational/neo-classical pedagogy to teach *about* the environment as part of a science curriculum. In turn, and as predicted by the notion of the duality of structure and agency, the conventions of science education encouraged those teachers to maintain their objective view of the environment through their practice of education *about* the environment (Giddens 1984).

The teachers who consistently objectified the environment incorporated SSP into a traditional science education for one of two main reasons: (i) in order to implement SSP as education *about* the environment, thereby supporting their personal anthropocentric environmental ideology; or (ii) as part of the science curriculum as directed by their principal. Whereas Simon chose to implement SSP as a science subject that reflected his strong personal environmental values, Lisa had been instructed by her principal to incorporate SSP into her usual science-based lessons.

Unlike Simon, Lisa indicated that her personal environmental values were not necessarily violated by her principal's directions to implement SSP, stating that she had "always been into the environment...and recycling" and that therefore "it's really good that the school is taking it [SSP] on". However, the principal and the SSP coordinator at South Bay Primary School indicated that Lisa was the only teacher at the school to have undertaken any university-level science learning, and was therefore responsible for the planning and the teaching of science throughout the entire school. She indicated that although "we've tried to integrate as much as possible, so we're mixing it [SSP] into maths, and reading and science", the majority of SSP-related activities were undertaken in science lessons, and that this was her responsibility. This seemed to contradict the intentions of Lisa's principal, Helen, who stated that her decision to implement SSP was to provide "teachers with a hook: a new way of teaching with a new way of learning", because she viewed SSP as a "vehicle for whole-school change". The underlying reasons for Helen's decision to give this responsibility to Lisa were not clear, and may have related to her understanding of the ability and/or willingness of other teachers at this school to engage with the program. However, irrespective of the reasons for Helen's decision, Lisa's pedagogical practice was strongly vocational/neo-classical (Fig. 5.5; Sect. 5.6.1); an unequivocally teacher-directed practice that contradicted all of the socially-critical goals of SSP. Thus, like Simon, Lisa implemented SSP as science education, and in so doing, adapted SSP to fit her understanding of how science should be taught, that is, her tacit knowledge of the practices of science (see Sect. 3.3). In turn, and as predicted by the notion of the duality of structure and agency (Giddens 1984), Lisa's tacit knowledge of the practices of science education encouraged her to maintain her objective view of the environment and her practice of education *about* the environment. The relationship of a teachers' tacit knowledge of educational practice to the development of rhetoric–reality gaps is discussed in Sect. 8.6.

Contrary to the practices of the teachers who consistently objectified the environment and implemented SSP through a vocational/neo-classical pedagogy, the teachers, namely Karen and Cathy, who referred to *our* environment, or, in relation to students, *their* environment, implemented SSP through a socially-critical pedagogy.

8.2.2 Our *Environment*

Karen and Cathy referred to the environment in a variety of ways, including as *our* environment, or in relation to students, as *their* environment. Both teachers positioned humans as custodians of the environment in phrases such as "we have not respected *our* environment" (Karen) and we must "make less of an impact on *our* environment" (Cathy). This was also evident in Cathy's comments regarding the educational outcomes of SSP for her students: "to understand *their* environment, to learn about *their* environment, to have respect for *their* environment, and to actually act on that, and therefore, in the long term make changes to the world—starting with *their* world".

Despite referring to the natural environment in this manner, neither Karen nor Cathy disagreed with the environmental ideologies held by the teachers, such as Lisa, who referred consistently to *the* environment. Both Karen and Cathy used the terms *our* and *their* environment to represent anthropocentric environmental values which positioned humans as the owners and controllers of natural environments. Karen, for example, indicated that her environmental ideology was not that dissimilar to Simon's belief that human needs must take precedence over the needs of the environment, stating that she aimed to ensure that her students understood that they "can learn and can make a difference without really…making huge changes to their lifestyle". However, unlike Simon, both Karen and Cathy reported that implementing SSP, through a socially-critical pedagogy, enabled them to embrace their personal environmental ideologies through their teaching role. This indicated that, in light of the fact that all of the teachers seemed to understand the environmental and educational goals of SSP, as outlined in SSP documents, the generally anthropocentric environmental ideologies held by Cathy and Karen differed to those held by teachers such as Lisa and Simon. This supports the notion that an individual's environmental ideology cannot be simply identified as purely anthropocentric or ecocentric. The environmental ideologies held by these teachers are better understood as relatively more, or less, ecocentric. In this case, the environmental ideologies held by Karen and Cathy were anthropocentric, but more ecocentric than those held by the other teachers. A detailed discussion of the presence and/or validity of a "two-factor ecocentric/anthropocentric structure of beliefs about the relations between people and their environment" (Amérigo et al. 2007, p. 102) is beyond the scope of this discussion, but represents an ongoing focus for research (e.g. Dunlop et al. 2000; Milfont and Duckitt 2004; Schultz 2001; Thompson and Barton 1994).

Unlike Simon's idea that he could implement SSP through a 'balanced' stance, both Cathy and Karen found ways in which to incorporate their ideas regarding environmental values into their classroom practices. Their approach reflected Kelly's (1986) notion of committed impartiality, described by Cotton (2006) as "taking a committed stance while remaining open to alternative views, and avoiding imposing values on the students" (p. 77). Karen, for example, described her approach to SSP as: "I am trying to teach the word respect…and with that word, everything in my opinion that's environmental comes under that banner". Karen indicated that focusing on respect was not only important in terms of achieving student learning outcomes, but also in terms of satisfying her personal teaching goals:

> we have not respected our environment and that's why we've dug ourselves a hole, mankind…the flora and the fauna…not just the big creatures of the planet…the smallest ones are critical in the cycle of life because if you break one chain you have repercussions, and that's what we've been doing…if I can impart that to children, well then I have made a difference.

Karen indicated that SSP provided an opportunity for her to not only make "a difference", but perhaps to also take responsibility for her own role in the detrimental effects of human–environment relationships. Similarly, Cathy indicated that SSP enabled her to justify her role as a teacher by demonstrating that she was at the "forefront" of addressing issues she considered to be important. She noted that her participation in the SSP professional development sessions had greatly affected her:

> I look at the world in a totally different way…turning off the lights and not having as long a shower…it really does make you more aware…it affects your whole life, I'm much more aware of environmental issues now than what I was before I did the program.

Thus, unlike Simon and Elizabeth, Karen and Cathy recognised that some of the values that they perceived as being embedded in the goals of SSP correlated closely to their personal environmental ideologies; there was a high degree of value "congruence" (Coburn 2004, p. 277). Congruence, that is, the notion that teachers are "more likely to engage with new ideas or approaches, depending on the degree to which they…find ways to connect them with their pre-existing beliefs" (p. 277) is discussed in relation to the development of educational rhetoric–reality gaps in Sect. 8.4.

The rhetoric of the teachers indicated that they held environmental ideologies which, although best described as anthropocentric, reflected a range of ideals that could be scaled as more, or less, ecocentric. The teachers who implemented SSP through a socially-critical pedagogy held the most ecocentric environmental ideologies, and justified their practices by relating their personal environmental ideologies to the environmental values they perceived to be embedded in the SSP goals, or by interpreting the goals of SSP in ways that matched their own values. For some teachers, these values were not unique to environmental education.

Despite the obvious, and intended, relationship of SSP with environmental concerns, some teachers related their implementation of the program to personal values other than those directly related to an environmental ideology. David for example, stated that his motivation for implementing SSP was first and foremost:

> altruistic…to leave the world a better place…to build better pillars of society…it means leaving the community in a better way than you found it, not just taking up the environment and wasting space or using up the air, but actually contributing, actually making a difference.

In other words, David considered SSP a vehicle through which he could develop students' sense of social responsibility rather than address specific environmental attitudes. It is important to note that although David did not employ a socially-critical pedagogy in the strictest sense, his motivation for implementing SSP was not inconsistent with the goals of the program, and was congruous with the aims of the liberal-progressive pedagogy he employed (see Fig. 5.3). Both the teachers and the principals correlated the implementation of SSP with the notion of 'social responsibility'; particularly as it related to the 'democratic equality' purposes of education (see Sects. 6.2.1 and 6.3.1). Philip, for example, stated that the implementation of SSP meant that at West Quay Primary School, "hopefully we're helping to produce responsible, ethical individuals".

The ability of the teachers and the principals to relate the goals of SSP to the anthropocentric notions of 'social responsibility' and 'democratic equality', or to specific values such as 'respect', indicated that the goals of SSP could be interpreted in ways that appealed to a variety of personal values. Each of the teachers and principals chose to interpret the notion of sustainability, as represented by the environmental education advocated by SSP, in terms of ecocentric values (the environment is preserved because of its instrumental value), or in terms of anthropocentric values (the environment is preserved for the central needs of humans; Vilkka 1997). Several of the teachers, including for example Cathy, Karen and Lisa, chose the latter interpretation of SSP goals. Thus, although a teacher's personal environmental ideology may have predicted the manner in which they chose to interpret the goals of SSP, it did not predict their ability to implement SSP through a socially-critical pedagogy, and therefore, did not adequately explain the development of the rhetoric–reality gaps. In light of this, it was important not just to consider how the principals and the teachers interpreted the goals of SSP, but to investigate how implementing SSP influenced the values, beliefs and or attitudes of the principals and the teachers.

8.3   SSP and Teachers' Lives

Giddens' (1984) notion of the duality of structure and agency suggested that not only would the principals' and the teachers' values influence the manner in which they implemented SSP, but that their interaction with SSP would, in turn, also influence their values. Helen, for example, attributed part of her decision to implement SSP at South Bay Primary School to her environmental ideology. She strongly identified with the ideals and goals of SSP which she described as future-oriented education which aimed to ensure that the natural environment and the human life it supports will still be "here in 100 years". However, she also reported that such a notion of sustainability "ties in with [my] own life" and that implementing SSP was "also about sustaining [my] own life". In other words, Helen acted on her personal environmental ideology. She aligned her role as an educator with her personal values by implementing SSP, and reported that her decision to implement SSP made her feel that she sustaining her own life.

The teachers' reflections of the SSP professional development sessions provided valuable insights into what they perceived to be the effects of their participation in the implementation of SSP. Cathy, for example, indicated that prior to the SSP professional development sessions:

> I used to hear it [environmental messages], yes I was concerned, but I wasn't at the forefront of doing something with it, I am now, and I think that's made a huge impact on my life personally, my life as a teacher and my family life.

Cathy noted that the initial professional development programs not only increased awareness of current environmental concerns, but also "had a huge impact on the staff as a whole…I think it made a huge impact when we did the…eco-footprint on each of the staff members". In other words, Cathy saw that SSP provided opportunities for teachers to incorporate their new understandings, developed as they explored their own human–environment relationships, into their professional lives. Both Karen and Cathy reported that they felt empowered by their participation in the SSP professional development sessions, and that they viewed SSP as an opportunity to address their growing awareness and concern for issues arising from current human–environment relationships. Karen, for example, stated that "I see the Sustainable Schools Program as…trying to empower teachers, or really teaching teachers to empower children". Neither of these teachers considered SSP a cure-all solution, but rather an opportunity for them to contribute positively to both humanity and the environment.

Although neither Lisa nor Elizabeth reported being personally affected by their participation in the SSP professional development sessions, other teachers reported that these sessions had generated a great deal of distress. This effect was felt most by Robyn and Anita. Robyn stated that although her principal had initially directed her participation in SSP, since the professional development sessions "I've started getting really worried about the world and everything and I just think ooh we're such wasters…I do want to stop this environmental problem that's happening at the moment". She believed that her awareness of environmental issues had dramatically increased since participating in SSP, and that "it's scary, and I know at home, even my habits at home have changed…I'm really conscious, for instance when you go out for dinner or…take away, how much rubbish you are using…it does hit home". Like Robyn, Anita noted that the "professional development sessions were…quite depressing at times" and that she had begun to question the ability of any one person to actually make a difference. Anita went on to explain that these feelings extended beyond the classroom, because they were so "personally overwhelming…sometimes you go to the supermarket and you see those people who get their plastic bags…you almost want to go and hit them over the head with something, and you sort of think…don't you get it?" She indicated that "I was really interested in doing it [SSP]…I'm interested in those kinds of things, and tried to put some of those practices into my own life, and…we're being so bombarded by media…you almost feel guilty if you're not trying".

Robyn and Anita's comments highlighted some unintended consequences of the SSP professional development sessions. The presenter from the Centre for Education and Research in Environmental Strategies (CERES) who had conducted many of these sessions was both surprised and dismayed by these reports. Although the professional development sessions aimed to establish a "language of possibility" (Fien 1993, p. 10), that is, the belief that each individual has opportunities for positive change (Wink 2000), these teachers' comments highlighted Giddens' (1979) understanding that any human action, irrespective of the underlying motivations, tacit understandings or deliberate planning, may create both intended and unintended consequences (see Sect. 3.3). In this case, the unintended consequences of feelings of helplessness and despair have been noted to occur in individuals as a response to their perception of a seemingly endless barrage of messages of environmental crises (Grun 1996).

Thus, the teachers responded to the professional development sessions in very different ways. Although it was not possible to establish a definitive cause and effect relationship, the teachers who successfully employed a socially-critical pedagogy to implement SSP, such as Karen and Cathy, did report feeling 'empowered' by the opportunity to address their personal concerns to minimise the effects of modern human–environment relationships. These teachers embraced a "language of possibility" in that they believed that SSP enabled them to make a positive contribution to society—a contribution that was congruous with their environmental ideologies. On the other hand, some of the teachers who chose to implement SSP through a vocational/neo-classical pedagogy, and whose practices defined a rhetoric–reality gap, indicated that the professional development sessions caused them to become anxious and guilty about the environmental effects of current human–environment relationships. Although these teachers acknowledged that their personal environmental ideologies were consistent with the values embedded in SSP goals, and recognised that these goals were best achieved through a socially-critical pedagogy, they were unable to act upon these understandings. As noted in Sect. 2.2 in relation to the development of environmental education "too much environmental knowledge (particularly relating to the various global crises) can be disempowering, without a deeper and broader learning process taking place" that enables individuals to respond, through action, to their developing awareness and understanding (Sterling 2003, p. 19). However, despite the fact that SSP provided opportunities for these teachers to act upon the new understandings gained from the professional development sessions, they failed to do so.

A detailed investigation of the degree to which the teachers' environmental ideologies were affected by their participation in the implementation of SSP, or the degree to which such effects enabled or constrained the teachers' ability to employ a socially-critical pedagogy, was not undertaken. However, in a study which compared the beliefs and practices of teachers in Hong Kong and England, Lee (1993) found that although teachers "espoused support for teaching attitudes of concern for the environment…[there was] little support for those teaching strategies that might enable teachers to achieve this aim" (Cotton 2006, p. 69). It was therefore important to investigate the relationship between the teachers' educational ideology and the manner in which they chose to implement SSP.

8.4   Educational Ideology

The rhetoric of the teachers (discussed in Chaps. 6 and 7) indicated that each teacher's beliefs influenced their perception of the enabling and constraining effects of the ontological elements of their work environment that they considered to be most critical to their ability to implement SSP. However, those beliefs did not adequately explain the development of the rhetoric–reality gaps observed in the implementation of SSP. Further consideration showed that, compared with teachers who implemented SSP through a vocational/neo-classical pedagogy, those who employed a socially-critical pedagogy held different educational ideologies, and therefore, different beliefs about their role as a teacher. This supports the notion that a teacher's practice reflects "not only the social and cultural context in which it takes place, but also individual considerations about what it means to be a teacher" (Michalak 2007, p. 77).

The classroom practices of teachers reflect a wide range of educational ideologies. The role of different educational ideologies, particularly in terms of their competing interests and relationship to the changing purposes of education, is the subject of ongoing discussion and debate (e.g. Goodwin 2007; Gray 2009; Harvey 2005; Kemmis et al. 1983; Spring 2004). However, as highlighted in Chap. 5 (Figs. 5.1, 5.2, 5.3, 5.4, 5.5 and 5.6), the educational ideologies central to understanding the practices of teachers, and the development of rhetoric–reality gaps associated with the implementation of SSP, were expressed through the predominantly vocational/neo-classical pedagogy employed by teachers such as Lisa and Elizabeth, the predominantly socially-critical pedagogy of teachers such as Cathy and Karen, and the many variations in between, such as the predominantly liberal-progressive pedagogy of teachers such as David and Julia. Each of these teachers' practices reflected, in part, a particular relationship between their personal educational ideology and that embedded in the socially-critical pedagogy advocated by SSP. As indicated by the principals (see Sect. 6.2.2), the decision to implement SSP was, most importantly, to encourage teachers who consistently practiced a vocational/neo-classical pedagogy to begin to use a socially-critical pedagogy. However, the process of changing to a socially-critical pedagogy "requires practitioners to reconceptualise their curriculum and to question prevailing practices. The issue is that a socially-critical environmental education does not cohere, in many cases, with practitioners' theories of teaching, learning and curriculum" (Walker 1997, p. 158). This was central to the development of the rhetoric–reality gaps in the implementation of SSP.

Each teacher's educational ideology reflected their beliefs about what they considered to be the most important aspects of their role as a teacher. These are discussed with respect to: student emotional wellbeing (Sect. 8.4.1); student ability (Sect. 8.4.2); student choice (Sect. 8.4.3); learning from others (Sect. 8.4.4); assessment of student learning (Sect. 8.4.5); and teacher power and classroom authority (Sect. 8.4.6).

8.4.1 Student Emotional Wellbeing

Some educational rhetoric–reality gaps have been attributed to teachers' beliefs that the proposed changes inhibited their ability to appropriately care for their students (Bailey 2000). When considering the implementation of SSP through a socially-critical pedagogy, comments from some of the teachers suggested that they were concerned about how the authentic learning experiences encouraged by this approach would affect their students' emotional wellbeing.

The notion that "emotions are at the heart of teaching" (Hargreaves 1998, p. 835) was highlighted by Cathy's comment that teachers must manage myriad emotional issues each day: "there's just so many things you have to cram into your day… [including] your emotional problems with kids, and your emotional problems with parents". Although Cathy's comments reflected the need for teachers to deal with their personal emotional wellbeing at school, teachers must also consider the emotional wellbeing of their students.

Managing student emotional wellbeing is, however, a contentious matter (Brunker 2007). The degree to which schools, and therefore teachers, are responsible for the development of students' emotional wellbeing, and the manner in which they should attempt to fulfil this responsibility, is a subject of continuing debate. Although Australian primary school principals are concerned about the ever-increasing demand for schools to accept a greater responsible for students' emotional development, particularly because "there were increasing numbers of children beginning school who lacked the necessary social and language skills and the ability to concentrate" (Angus et al. 2007, p. 2), many researchers have shown that improving student wellbeing leads to improved academic achievement (e.g. Brunker 2007; Caprara et al. 2000; Fook et al. 2005; Malecki and Elliott 2002). Many aspects of the socially-critical pedagogy advocated by SSP incorporated practices consistent with practices that are also associated with improving aspects of student wellbeing. Brunker (2007), for example, reported that student wellbeing, or "social emotional wellbeing" (p. 2) is best developed through schooling which enables the students to develop and to participate in meaningful relationships with their peers, their teachers and their community (Brandt 2003). She noted that all aspects of a learning environment can impact on student emotional development, stating that "it is crucial that everyone connected to schools recognise this role in order to ensure that both their explicit and implicit behaviours enable children to develop and experience positive social emotional wellbeing" (Brunker 2007, p. 2). However, some of the teachers found that their interpretation of this responsibility constrained the manner in which they could implement SSP because they believed that some of the authentic learning experiences encouraged by a socially-critical pedagogy inappropriately compromised their students' emotional wellbeing. This concern was best illustrated by the differences in the manner in which Elizabeth and Karen dealt with the topic of death with their students.

Elizabeth reported that the teachers at Mountain Primary School carefully managed the type of information shared with students, particularly regarding real life events in the kitchen garden. For example, the horticultural manager of the kitchen garden (Stephanie) explained that when a student found an injured bird in the school grounds and brought it to her, she took the bird and placed it out of sight until it died. She then informed students that the bird had recovered and flown away. Neither Elizabeth nor Stephanie believed that primary school students should be exposed to the topic of death. In contrast, Karen explained that all of her students, irrespective of age, should have opportunities to be engaged in authentic, real world learning, and that this meant that her students should not shielded from the natural cycle of life and death. The following passage highlights the view of life obtained by groups of students involved in the artificial breeding program for an endangered species of bird at East Valley Nature Park (EVNP). Karen described events related to the first few chicks to hatch:

They [the park rangers] take it [a chick] away from the parent, and they explain why they take it…they don't handle them unless they absolutely have to…they got an ostrich feather duster, and so if it was scared it would go under the duster…that went on for weeks and weeks and weeks and weeks…until one day…the little guy had got his neck caught…and had strangled itself to death…terrible thing…because we all know they're critically endangered and it's taken two years to actually get a live one…and then we said…let's not despair, we've got a couple of eggs…the next thing you know, the next egg has hatched at the wrong time…it died in the egg…[and there has] just been this whole succession of disasters, but the kids are aware of them.

Karen indicated that her role as a teacher was not to shield her students from these types of life experiences, but to ensure that experiences were handled in an age appropriate manner. She explained that:

we have [real world] activities but they're adapted according to the age [of students], for instance…the artificial insemination [as part of an artificial breeding program], you probably wouldn't talk about that a real lot with the real little ones because they just simply wouldn't comprehend, but the older ones…it's really powerful stuff.

Karen's willingness to expose her students to the real world enabled her to share and discuss such events openly with students, and encouraged students to critically reflect on their experiences in a manner consistent with the goals of a socially-critical pedagogy. This was demonstrated by the students' questions and discussions with a visiting bee keeper. One Grade 3 student, for example, enquired if bees were able to control their rate of breeding in order to respond to the amount of food available in any season. This student explained that her question was prompted by what she had learned about the way in which kangaroos controlled their breeding, through assisting in a kangaroo monitoring program at EVNP. This prompted a class discussion that focused on the complex ways in which human activities (both good and bad), climate change and natural seasonal and species' life cycles interrelate, and how in turn, these relationships might influence not only the bee keeper's business, but also the lives of the students' families. Despite the natural hurdles and setbacks experienced by the students involved in various programs at EVNP, Karen noted that the students felt connected to the park, highly motivated to return, extremely proud of their contribution, and most importantly, highly comfortable with their ability to influence the world in ways they considered to be positive.

The socially-critical pedagogy advocated by SSP incorporated opportunities for students to learn through their participation in authentic life experiences, and therefore also presented students with opportunities to learn how to deal with the reality of setbacks. Some of the teachers viewed such setbacks as learning opportunities, whereas others considered even the most innocuous mistakes to be an unacceptable component of student learning. This was well demonstrated by the different approaches to learning activities taken by Lisa and Cathy. As described in Sect. 5.6.1, Lisa carefully controlled every aspect of her science classes. Lisa ensured that each student obtained the information that she considered to be most important by directing the class to carefully and accurately copy her data from the whiteboard into their work books. The only problem-solving aspect of this lesson (determining the percentage of water lost from fruit that had been left outside in the sun) was directed by Lisa so that all of her students followed very precise, step-by-step instructions. There was no opportunity for her students to estimate or guess, or make a mistake of any kind. There was no opportunity for her students to question, or to "discover" anything for themselves. In other words, the students in Lisa's class were not given the opportunity to experience failure.

In contrast, Cathy described a project in which her students worked in small groups to design and build a racing car to be powered by a hydrogen fuel cell:

the whole unit was the mistakes that the kids made...occasionally they were disappointed about it, but the learning that went on from the mistakes that they made, and saying okay this didn't work, what can we try...I reckon the beauty of that [was] it was just open, other than giving them the ultimate [answer, we gave them] what we wanted them to try and achieve...and to see the way that they attacked that and the testing and investigating that they did.

Cathy explained:

I mean there are times, okay when you want them to learn particular skills then you teach them that skill, but if it's just something that you want them to learn about a particular topic, I think they have to make those mistakes, and sometimes, well no, I don't think we had one child that didn't have failures during that [project], no one was really cut about it.

Cathy's approach showed that she valued the learning that came from the experience of failure. Her comments indicated that setbacks encouraged her students to question their ideas and to think creatively in order to find solutions to the problems that they encountered. The students considered the results of the car races at the end of the project to be secondary to the experience of undertaking the project. The students were most keen to explain the design of their cars in terms of: the testing they had undertaken; the research they had conducted into the properties of different materials; and what they had discovered about the performance of different design features, particularly with respect to size, weight and friction. It was evident that the students had developed a wide range of understandings through their participation and negotiation in a collaborative and cooperative work environment.

8.4.2 Student Ability

When investigating the educational benefits of programs in which teachers collaborated with science experts, Trautmann and MacKinster (2005) found that the "teachers' perceptions of their students' expectations and abilities" were a significant hurdle to educational change (p. 1). Similarly, some of the teachers who were required to implement SSP believed that their primary school students were too young to benefit from the learning experiences provided by a socially-critical pedagogy, a feeling epitomised by David's statement that the practice of socially-critical pedagogy concerned him because "the kids just I don't think are capable".

Elizabeth, for example, held strong beliefs about the ability and capabilities of her primary school students. When commenting on a socially-critical pedagogy described in a hypothetical scenario (see Table 4.2) in which students visited a local forest to learn about plants in order to inform their decisions when planning a native garden for their school environment, Elizabeth described the learning experiences as more suitable for "secondary school, or later". She suggested that such an approach "would fit in very well with the sorts of subjects they [secondary students] do, the way that some of those subjects are run, it would be something that would be very easily managed in the secondary surroundings" and that "you wouldn't have [students from] primary schools going out into the community". These comments reflected Elizabeth's belief that her students were too young to benefit from the types of learning experiences potentially provided by a socially-critical pedagogy. In support of this, Elizabeth indicated that she did not believe that her young students were capable of completing even a well-practiced routine, such as the rubbish management system that they undertook each week: "because the kids can't do it unsupervised [it] takes up 20 minutes minimum". When questioned about why different classes in the school did not share the responsibility for waste management, Elizabeth replied:

> for example if you had one person looking after junior bins and one person looking after middle school bins and likewise for the senior school you'd still have to have someone overseeing it, and if you happen to have two of those supervising teachers away on one day, then you'd have to go through the explanations with CRT's [casual relief teachers], and that would be horrific.

According to Elizabeth, her students could neither carry out the rubbish management routine by themselves, nor were they able to explain the requirements to a visiting teacher. This seemed to contradict Elizabeth's own views of how to determine if learning had occurred:

> it's not until you can tell or teach or impart that knowledge that you actually know it yourself, and that you know you know it, so when I see them [the students] imparting it to someone else…whether or not it's someone else in our grade or another grade, or sharing that information…that's when I would know the extent of the learning and the extent of the success of that aspect.

Elizabeth's comments provided a valuable example of the effect of Giddens' (1984) notion of the duality of structure and agency in the classroom. Each time Elizabeth directed the students' implementation of the rubbish management system, she re-confirmed that the rubbish management system was a routine of moving rubbish bins, and that part of this routine was that students are incapable of undertaking such a routine on their own initiative. As Elizabeth never provided her students with the opportunity to demonstrate that they did have the ability to manage this program, she never observed the evidence she required to indicate that they did have this ability. In turn, Elizabeth's belief that her students did not have the ability to undertake the rubbish management system was never challenged. Thus, as discussed in Sect. 7.2, Elizabeth's practice was highly routinised, and as demonstrated here, her routinised practices incorporated her beliefs about her students' abilities.

However, contrary to Elizabeth's belief that primary school students were unable to take responsibility for almost any aspect of their education, students of a similar age in both Karen's and Cathy's classes appropriately and very effectively directed many aspects of their learning. However, both Karen and Cathy carefully and deliberately considered the age of their students when facilitating opportunities for student learning. Karen, for example, indicated that the appropriateness and effectiveness of any learning activity "really does depend on the age of the children as to how sophisticated their skills are or whether a child could actually do that or not". Both Karen and Cathy encouraged their students to make decisions not only about their how they demonstrated and extended their understandings after participating in a learning activity, but also about what learning activities to undertake. In other words, they expected students to make decisions regarding their learning, and they respected the decisions made by students. Karen and Cathy considered that part of their role as a teacher was to assist students to learn how to make choices; a central component of any effective socially-critical pedagogy in which teachers "are not seeking right answers, but engaging our students in a process that can help them to make better decisions" (Jickling 2005, p. 43).

8.4.3 Student Choice

Elizabeth described the role of student choice in normal classroom activities as important, or "good, as long as they [students] know what choices they've got". This suggested that Elizabeth viewed student choice to be a teacher-directed activity. Similarly, Lisa could provide only one example of a situation in which her students had been given an opportunity to make a decision, related to "our science and maths night coming up, where the kids have decided what experiments and what activities [of those already completed in lessons] they want to show their parents when they come in". In other words, both Lisa and Elizabeth tightly controlled student choice and limited their students' decisions by defining specific alternatives.

In contrast, both Cathy and Karen indicated that student choice was an integral component of their pedagogical approach to SSP. Cathy stated that "it's really important to give the kids a choice of ways of presenting their information [to] allow for the different learning styles" and commented that students "come up with such brilliant things to do". Karen suggested that this was also important because "how they [the students] come back from that [any learning experience] is really a personal thing" and that it was "up to them as to where they take, where they go with the information that they've been given" and the ideas they developed. Karen explained that "I find every week, of every group [of students] that come here, we go in a different [direction]—even though we have these basic...components that we will look at for the week, we all end up going in a different way". She stated that at EVNP, "I think there's a fair degree of freedom here in choice" and cited a project in which students of different ages assisted in a biological survey of a pond at EVNP as an example of how the students direct their learning:

> some will come back and want to really go on with the microscope...others won't be interested, they'll want to come back and they'll take cameras and they'll want to do a movie

about it…others will just want to work with photographs and maybe do a standard…slide show type of thing.

Both Cathy and Karen understood that different students would learn different things from the same activity, but that for each student, the learning would be valid. In other words, these teachers believed that valid learning at school incorporated more than just the learning able to be identified or directed by a teacher (see Sect. 8.4.4). Cathy and Karen agreed that student choice was important, not just in terms of developing their students' ability to make appropriate decisions, but most importantly, also to motivate their students. Cathy explained that her approach to student choice reflected her understanding that "I can't see that you can get excited about something that you've done twenty times before, that's not what I would be after… [I need] to consult [with] the kids…[ask] where they want to go". She referred to a student initiative to establish a "leadership role" in their school by advertising the environmental benefits of bringing no-rubbish lunches, as a project in which student choice had led to learning activities that had motivated her students to excel. Cathy stated that because the students had "planned the whole thing", including the development of a comical advertising character (the 'Nude Food Dude') and the use of slogans such as "eating nude food makes you a cool dude", they had been highly motivated to succeed, and deeply engaged in the new and sometimes unusual learning activities they designed.

Similarly, Karen explained that there were opportunities for her students to participate in many types of learning activities at EVNP, but that many of these activities would begin only after she had identified a group of students who were enthusiastic and willing to participate. For example, she explained that a worm farm had been donated to the school, but that:

we haven't started because I need a grade that needs a bit of a ground swell, okay, so I'm showing the kids where it is, and we have a little bit of a talk about it, and they've got to be a group who are really keen and really interested…a couple of groups will have real ownership of that and others, whilst they might participate and learn it, they won't have that same ownership…I don't expect children to all have a high level of ownership for everything they do, that's just ridiculous.

Karen's comment indicated that she acknowledged that all students are individuals with a range of interests, and that any student will learn best when they are interested in the learning activities in which they participate. In other words, Karen, and Cathy, believed that by encouraging their students to make choices about their learning, their students were able to identify those aspects of an activity which interested them the most, which in turn, motivated them to more deeply engage with that learning activity; like Richmond (1990), they understood that "the link between motivation and learning is strong" (1990, p. 194). However, these teachers did not completely relinquish their decision making role, as indicated by Karen's comment that "we do need to have teacher direction and control", but instead, moderated their decision making in recognition that "it's just simply not as powerful as coming from the children".

Lisa also recognised the role of student interest. Despite limiting student choice through a strongly vocational/neo-classical pedagogy, Lisa indicated that student interest was an important consideration for teachers. She explained that the implementation of SSP had improved her pedagogy by encouraging her to incorporate more experiments in her science classes, stating that this was beneficial as:

> it's given that hands-on feel to it [science]…they [the students] like it, they enjoy it, they want to be part of it, they wanna know what it's about, so that's why we do the hands-on experiments because it really draws them in.

Lisa also described her observation of the responses of her students to an activity run by an external educator who took them for a walk along the banks of a local river to investigate water pollution issues. She noted that:

> the kids loved that [be]cause we were there, and they could see it, and they knew what we were talking about, and they were telling us things, [be]cause that was their area and they know what happens and they play there.

Lisa's comments demonstrated that she was aware of the benefits of student interest in learning activities. However, Lisa's teacher-directed pedagogy suggested that she believed that it was a teacher's role to identify ways in which to make lessons interesting. In other words, Lisa believed that a teacher must consider student interest when choosing the content of a lesson and the manner in which that content should be learned.

In contrast, Elizabeth seemed unable to identify any aspect of her role as a teacher which required her to consider student interest. Her position on this was best illustrated by comments related to her use of hands-on activities:

> hands-on is definitely good…I mean I'm thinking about our students who, many of them are from a low socio-economic level…often you've got to, you know, they don't get past 'I like doing that'…but why? or what did you learn? Well, you know, it's limited feedback… hands-on [activities] and working with a partner is a really good way of drawing out and pushing out and maximising their [the student's] own understanding of some learning.

This comment illustrated Elizabeth's belief that her role as a teacher was to ensure that her students acquired certain knowledge. When commenting about a hypothetical scenario that depicted a vocational/neo-classical pedagogy (see Table 4.3), Elizabeth noted that "the teachers have looked at the way that those students engage best". In contrast, she noted that the socially-critical approach represented in another hypothetical scenario (see Table 4.2) "is not quite as thorough in looking at how to engage most of the students for most of the time". These comments clearly indicated that Elizabeth believed that her students were not capable of making decisions regarding their learning, and that her role as the teacher was to identify the lesson content and learning procedures that would best assist her students to acquire specific knowledge. Although Elizabeth's comments indicated that, to some degree, she recognised that learning was probably most effective when the students were engaged in the learning process; this interest was secondary to the need to design lessons which addressed the knowledge that she had determined the students must learn.

Thus, although most of the teachers indicated that student learning was most effective when the students were motivated to learn, and that this motivation was derived from an interest in the learning activities, the manner in which the teachers acted on this understanding differed greatly. Both Cathy and Karen believed that student interest was paramount, and that their teaching role was to motivate their students by facilitating a wide range of learning experiences from which their students could choose to participate. These teachers viewed the socially-critical pedagogy advocated by SSP as a vehicle through which they could assist their students to develop the skills and understandings that they needed to make the most of any learning opportunity. Many of these learning opportunities incorporated activities that were undertaken within, and which contributed to, real world contexts. Cathy and Karen demonstrated that primary school students were capable of directing many aspects of their learning, both in terms of choosing learning activities, as well as choosing ways in which to demonstrate their learning.

In contrast, Lisa and Elizabeth aimed to develop student interest through a vocational/neo-classical pedagogy. When planning learning units, these teachers aimed to identify topics that they perceived would interest the majority of their students. They aimed to promote student motivation and interest by incorporating opportunities for their students to choose, from a set of appropriate pre-planned alternatives, some aspects of some of their lessons. In other words, while Cathy and Karen focused on facilitating learning activities driven by student interest, or student choice, Lisa and Elizabeth focused on attempting to make their students interested in the teacher chosen, pre-planned, learning activities. Neither Lisa nor Elizabeth believed that primary school students were capable of directing any significant aspect of their learning, and therefore, tightly controlled the learning process in a manner that prevented their students from demonstrating their ability.

Thus, the practices of teachers such as Lisa, Elizabeth, Cathy and Karen demonstrated that the teachers' beliefs about the ability of their students, in part, reflected those teachers' observations of the actions of their students. In each case however, the actions of their students also reflected the practices of their teachers. This demonstrated the influence of the duality of structure and agency (Giddens 1984) on the teachers' agency. Richmond (1990) noted that the "key here is the probability that motivated behaviour will occur regardless of the presence of others, whereas the compliant behaviour will only occur in the presence (physical and/or psychological) of the compliance-seeking person" (p. 182). Richmond (1990) suggested that these two effects reflect two distinct classroom management styles: one which aims to motivate students, as undertaken by Cathy and Karen; and one which concentrates on achieving student compliance, as undertaken by Lisa and Elizabeth. In other words, Cathy and Karen taught in a manner that enabled their students to capably direct aspects of their learning, which in turn, demonstrated that their socially-critical pedagogy was effective. Lisa and Elizabeth taught in a manner that prevented students from undertaking any decision making regarding their learning, which in turn, prevented students from demonstrating that they could effectively learn from experiences potentially provided by a socially-critical pedagogy.

The comparison of the teachers' practices and consideration of the teachers' rhetoric about those practices indicated that the educational ideologies held by the teachers who aimed to motivate students were somewhat different to the educational ideologies held by those who aimed to achieve student compliance. One of these differences concerned whether or not a teacher was the sole source of knowledge in the classroom. In other words, some of the teachers believed that the process of learning was not restricted to teacher–student interactions.

8.4.4 Learning from Others

Both Cathy and Karen believed that an important part of assisting their students to develop the skills and understandings required to take advantage of potential learning opportunities, and therefore to motivate them as learners and help them to identify their interests, was to encourage the development of meaningful relationships. Meaningful relationships included those between peers, as well as those between students and members of their community. Cathy, for example, stated that "we've had a lot of interaction" with the local community. She listed a wide range of collaborative projects in which her students had participated, including: the production of public artwork in the form of illustrated poems and ceramics as part of an upgrade to a local park and playground; the development of a local history path; and a variety of community planting projects and pollution monitoring programs. With respect to her students Cathy noted that "I think it's great that they're working with other people in their community…I think it gives the kids a really good sense of community and how they can be part of it, and how different members of the community can contribute".

Similarly, Karen felt that her students were highly engaged with their experiences at EVNP not just because of their ability to make decisions about their learning, but also because of the quality of the relationships they developed with the staff, resident animals, park visitors and natural environments. She stated that "I try to make it so that these are their [the students'] endangered birds and these are their eggs…these are their ducks…these are their lizards, so they have ongoing relationships". Karen explained that, in response to student requests, she had established a postal service which enabled students to maintain communication with the EVNP staff between visits:

> I'm the postie, the mail lady, and so there's the EVNP letter box…in any one day I bring the mail and I drop it off to the rangers who then send the letters back, and there's a heavy flow of writing letters…from the children to the rangers and back and forth and back and forth… for instance a grade writes a letter [saying] "we saw a bird it looks like this" and they draw it, and so the guys [EVNP staff] write back to them and say "it's probably such and such… if you want to come up I can show you some other examples of that species".

Karen noted that the effects of the development of such relationships extended beyond school hours, indicated by the fact that "so many kids come up here at the weekend and out of school time". She explained that her students were often seen

on weekends at the park either explaining the importance of the endangered breeding program to public visitors, or chastising visitors for dropping their rubbish in the
park. Karen stated that the relationships developed by the students at EVNP were
valuable as they represented "the real world" and that this meant that the students
were learning in an environment that required them to develop their understanding
of "team work". She believed that this ensured that the students "can learn and can
make a difference" and most importantly, that "children get the message not just
from a teacher but from other people" and this makes their learning "really powerful" because it has real "meaning".

These comments suggested that meaningful learning is knowledge and understanding that students can identify as being relevant to their lives, and relevant to the
activities in which they are motivated to participate. This reflects the understandings
upon which a socially-critical pedagogy was founded, as indicated in Sect. 2.3.2,
that learning is only truly effective when developed within the contexts related to a
students' life experiences (Giroux 1988), that is, within their "community"
(Mogensen 1997, p. 434). Similarly, by undertaking a socially-critical approach
both Cathy and Karen acknowledged Freire's (1972) understanding that learning
opportunities which incorporate contextually specific experiences in this manner
positions each student as an "active actor" (Swain 2005, p. 1). Freire (1972) believed
that this was central to an effective socially-critical pedagogy, because it provided
opportunities for students to not only increase their awareness of the world, but to
also purposively critically reflect on their developing understandings through
authentic participation (Schugurensky 2002, p. 63).

Thus, the vocational/neo-classical pedagogy employed by Lisa and Elizabeth
reflected their belief that their students were not capable of directing any significant
aspect of their learning. As a result, these teachers assumed a role in which they
judiciously identified, planned and prepared all learning activities in which their
students would participate, and from which they would learn the knowledge identified by their teachers as appropriate. In contrast, the socially-critical pedagogy
employed by both Cathy and Karen demonstrated their belief that their students
were capable of making appropriate decisions regarding their learning, and able to
effectively participate in, contribute to, and learn from, authentic community-based
activities. The beliefs held by these teachers regarding the ability of their students
clearly influenced the manner in which they undertook their role to: care for the
emotional wellbeing of their students; identify appropriate learning; and facilitate
student motivation and interest. These beliefs influenced each teacher's decision
regarding whether or not to implement SSP through the recommended socially-
critical pedagogy, and therefore, contributed to the development of the educational
rhetoric–reality gaps identified. However, other beliefs also contributed to the development of these rhetoric–reality gaps. These beliefs centred on the problem of
defining, recognising and assessing valid learning.

8.4.5 Assessment of Student Learning

Cathy believed that part of her role as a teacher was to motivate students: "I have to be motivated to get the kids motivated". She suggested that sometimes this meant learning alongside her students. She used the example of a recent project: "I knew nothing about that [project] before we started, [but I] know lots about it now". This reflected the idea that when students are able to make decisions they may choose to undertake a project about which a teacher knows nothing, and this makes assessment of student learning somewhat difficult.

Similarly, Karen noted that, although the learning experiences at EVNP had real meaning for her students and that undertaking any of the activities meant that her students were "using the whole curriculum", it was difficult to assess that learning in terms of a traditional tick-the-box curriculum. Scott and Oulton (1999) also reached this conclusion, and suggested that the environmental education advocated by socially-critical approaches, like that embraced by SSP, are not often successfully implemented as "school success continues to be measured in terms of traditional academic, rather than more-environmental, criteria" (p. 90). Karen reported that at EVNP "we sit down and say right, where do we want these kids to be and we actually create the curriculum". This so-called "curriculum" incorporated the types of experiences and activities available to the students at the park rather than simply listing specific outcomes in terms of knowledge gained. She noted that this did not in any way preclude the teaching of specific ideas and/or skills, noting that particularly in her role as an Information and Communication Technology (ICT) teacher, "at the end of the day you do still have to teach some skills, because you're not going to get these final whiz bang products unless the kids have been taught the skills in the first place". Karen indicated that activities undertaken by her students at EVNP constituted such rich learning experiences that she could easily justify these in terms of any standard curriculum if necessary: "I can say that without even reading it [the state government curriculum]" but that, "at the end of the day yes, we do need to assess [in relation to the curriculum], painful as it is". However, the idea that a teacher should endorse a learning activity without first deciding what should be learned from that activity, and how that learning reflected a specific component of an appropriate curriculum, was not acceptable to most of the teachers.

David indicated that a vocational/neo-classical pedagogy was preferable to a socially-critical pedagogy because it enabled a teacher to fulfil their role to "lead kids". He indicated this meant that, as a teacher, "you're giving them [students]... everything they need...you're putting it in a bit of a framework". David suggested that when "the range of things that could be done was fairly certain...the outcome was fairly certain" and ensured that the students were "getting definite outcomes in the curriculum". David's comments suggested that he linked student decision making with his need as a teacher to demonstrate that his students had accomplished appropriate curriculum outcomes. Both Elizabeth and Lisa shared these sentiments.

Elizabeth seemed to have significant difficulty stating what she perceived to be the goals of SSP, and consistently correlated the SSP goals with the transfer and acquisition of knowledge as represented in the state government curriculum. Elizabeth held very clear views on the "correct" way in which to teach. She described a vocational/neo-classical pedagogy as "a very logical and progressive approach", and that one of the best ways in which to monitor learning was to observe students "imparting it [knowledge] to someone else" because the ability to "repeat and teach someone else about it…or impart that knowledge" is the only way in which to "actually know it yourself, and that you know you know it". These comments highlighted Elizabeth's belief that recitation of factual knowledge was an appropriate assessment of student learning, because "that's when I would know the extent of the learning and the extent of the success of that aspect [of a teaching/learning unit]".

Like Elizabeth, Lisa indicated that her role as the teacher was to determine what students must learn, stating that "I want to give them [the students] the education". Her lessons demonstrated that she aimed for each student to acquire what she had determined to be appropriate knowledge, and that this knowledge was related to a specific curriculum. Lisa's comment that "at the moment most of what we do is just the curriculum side of things" reflected her attempts to implement SSP, as a new curriculum, by incorporating new and/or different knowledge into her lessons.

Elizabeth and Lisa shared the belief that a teacher's role was to ensure that their students gained the knowledge outlined in a chosen curriculum, the success of which could be judged by testing their students' knowledge. These beliefs corresponded to the principles and educational aims of the vocational/neo-classical pedagogy (Kemmis et al. 1983) through which these teachers chose to implement SSP. Research by Trautmann and MacKinster (2005), for example, showed that teachers with these beliefs can find it difficult to embrace non vocational/neo-classical pedagogies. They found that attempts to introduce inquiry-style science lessons are often not successful because "many teachers view factual knowledge as the most important student outcome" of school learning (p. 2). However, it is important to note that the teachers who implemented SSP through a socially-critical pedagogy did not assume that factual knowledge was unimportant, nor did they refrain from assessing student learning.

Both Cathy and Karen agreed that assessing student learning from a socially-critical pedagogy presented some unique challenges, particularly because the SSP goals were best understood as student actions or behavioural change. Measuring these types of outcomes can be difficult, as stated by Karen:

> this program [SSP] is not meant to be an end in itself, it's meant to be a springboard [for] changing attitudes for children to take back to East Valley Primary School…now I don't know what I can truly measure here.

Karen referred to the actions of her students as evidence of their learning at EVNP. In addition to student actions described as evidence of motivation and interest (as discussed in Sect. 8.4.3) Karen also gave the example of a student initiative in which money was raised to re-design a bird breeding pond:

[the] junior school council got together, and this totally came from the children…normally you make money in junior school council and you give it to this group or that group, they wanted it to go to the Musk Duck so they could breed, [it's] really really important that these ducks get together to breed.

Karen noted that "a lot of people wouldn't be happy because it is loose", that is, because these outcomes are not easily related to traditional tick-the-box knowledge-based learning outcomes. Similarly, Cathy noted that she assessed the success of the program through changes she observed in student behaviour, citing for example:

it's amazing now you'll go out of the classroom for a specialist session, and in the middle of winter they'll turn off the lights on you, you can be still working there, but they've walked out of the classroom and turned off the lights.

Cathy stated that she had heard from both parents and students that "they've [the students] introduced lots of the things that we've introduced at school at home, so it's making an impact on the various home lives as well". Cathy indicated that this was evidence of learning.

Thus, both Cathy and Karen recognised that their students were transforming their developing understandings into action, and that such voluntary actions were evidence that their teaching strategy and implementation of SSP had been successful. In contrast, Lisa and Elizabeth neither acknowledged these types of outcomes as representative of valid learning, nor did they believe that their students had the ability to voluntarily act in these ways (see Sect. 8.4.4). In light of this, it is evident that Lisa's and Elizabeth's beliefs about what constitutes valid learning, and how this learning is assessed, contributed to the rhetoric–reality gaps represented by their practices.

8.4.6 Teacher Power and Classroom Authority

Giddens (1979) considered power to be a "capability" (p. 68) or "transformative capacity" (p. 88) in that it reflects a person's ability to achieve specific outcomes from their actions. As described in Sect. 3.7, according to this definition, power is derived from the complex and dynamic interrelationship between contextually-specific rules and resources, and an individual's ability to exploit and mobilise these in order to create an asymmetric distribution of resources. In line with Giddens' (1984) notion of the duality of structure and agency, rules and resources also combine to mediate human interaction by defining social expectations for: behaviour; shared meanings for communication; and appropriate sanctions for non-conformity. These in turn identify the relative power, or domination, of certain individuals in social interactions (Turner 2003). Although the role of power is not a specific focus of this discussion, some aspects of the teachers' attempts to implement SSP did reflect the effects of the power held by the principals and the teachers.

Giddens' (1984) suggested that power structures are not absolute, and that therefore the principals and the teachers must be considered to be active human agents with the ability to influence and transform traditional patterns of social interactions within their school (Devine 2000). The interactions of the principals with the teachers, and the teachers with other teachers in the schools where SSP was being implemented supported this idea.

As explained in Sect. 6.2.2, the principals were responsible for the decision to implement SSP within their schools. Many of the teachers indicated that their decision to participate in the implementation of SSP was made only because they recognised the power held by their principal in this matter. However, other research has found that the power of school principals is not absolute. Evans' (1987), for example, employed a structuration research framework to investigate the attempts of a new school principal to alter some of the well-established teaching practices in a school. Although the assumption that, irrespective of the educational policies of the principal, the teachers were able to maintain relative autonomy in choosing their classroom pedagogical practices was partly supported by the research findings, some unexpected features of the ways in which power relationships worked within the school were revealed. In particular, the teachers who provided resources, in the form of knowledge and support to the new principal, were most able to influence that principal's decisions. The change in principal was eventually accompanied also by a change in the cohort of teachers at the school, because new teachers were employed according to their willingness and ability to implement new policies, that is, policies developed under the influence of the teachers who were supportive of the new principal. In addition, newly appointed teachers enjoyed the principal's support, and therefore also influenced the principal's decisions (B. Evans 1987). Several aspects of the manner in which the principals and the teachers responded to the implementation of SSP supported Evans' (1987) findings.

Philip's experience as the principal aiming to implement SSP at West Quay Primary School correlated well with Evans' (1987) findings (see Sect. 5.7). For example, Philip stated that it had been very difficult to convince the teachers to alter their well-established pedagogies, and that as a result, "when we're employing, we'll be looking specifically for specific skills, people who are…into curriculum development and those sorts of things". In addition, he noted that although he could not always influence the classroom practices of his teachers, he aimed to "try and match up the skills you have [in terms of teacher ability]…to make things flow better", and referred to the fact that one teacher:

> who initially came back [from an SSP professional development session] and threw cold water on the idea of sustainability ceased to be the middle years contact person [be]cause I thought, well, this is something I really want to push, and I can't have someone who's ultra negative about it being the main contact [teacher].

In other words, Philip acknowledged that his power was not absolute. Although he could influence many things in the school, the manner in which the teachers chose to implement his requests represented their autonomy in their classrooms.

In all cases, the teachers chose the manner in which they would implement SSP. Teachers such as Cathy and Karen for example, successfully fulfilled their principals' requests to implement SSP through a socially-critical pedagogy, because the environmental and educational ideologies embedded within the program correlated with their personal ideals. Both of these teachers reported that they received generous support from their principal. Other teachers, including Lisa, Julia and Anita, also received support from their principals, but, due to the lack of congruence between SSP and their personal environmental and educational ideologies, they chose to fit the new program into their existing practices. As noted in Sect. 6.5.1, Julia and Anita reported that the support provided by a principal could have significant ramifications. They noted that their principal offered training and mentoring to any teacher who indicated that they were willing to attempt to alter their pedagogy in order to effectively implement SSP. In other words, the principal acted to enable teachers to embrace change. However, colleagues who resisted change viewed the acceptance of this support by those teachers as a betrayal. This meant that the teachers who resisted change were denied assistance that enabled change, which in turn, established a structural impediment to change.

Thus, as indicated by Evans (1987), irrespective of a principal's directions, the teachers most resistant to making the changes required to implement SSP through a socially-critical pedagogy, such as Elizabeth, successfully used the power available to them to maintain their well-established classroom practices. Other teachers, such as Lisa, used the power available to them to implement only the aspects of SSP that they considered to be consistent with their personal ideologies. The teachers who supported the ideologies embedded within SSP, such as Cathy and Karen, used their power to holistically implement the program. This supported the notion that the teachers' classroom practices reflected those teachers' choices, or agency. In other words, the gaps between the reality of teachers' classroom practices and the rhetoric of SSP were a matter of teacher agency.

The manner in which teachers establish and utilise their power in the classroom has been extensively studied (e.g. Kearney et al. 1984, 1985; McCroskey et al. 1985; Richmond and McCroskey 1984, 1992). In general, researchers agree that the power wielded by a teacher is not developed unilaterally, but is "most essentially a form of professional authority granted by students who affirm the teacher's expertise, self-confidence, and belief in the importance of his or her work" (Vander Staay et al. 2009, p. 262). Thus, compared with the socially-critical pedagogy undertaken by Cathy and Karen, the vocational/neo-classical pedagogy employed by Lisa and Elizabeth did not reflect simply a difference in the matter of who held the greatest power in the classroom, but rather a difference in the manner in which power was utilised, or rather, how teacher authority was established in the classroom (Phillips Manke 1997). Both Cathy and Karen established an authority which enabled them to provide learning opportunities that they believed would both motivate and interest their students to develop their understanding of the world, and learn to act on those understandings. On the other hand, Lisa and Elizabeth established their authority to provide learning opportunities that would best motivate and interest their students, to gain the essential knowledge-based understandings and skills outlined in the government authorised curriculum.

However, irrespective of the manner in which the teachers' chose to utilise their power in their classroom, their classroom practices closely reflected their personal educational ideology. Carrington (2010) suggested that any teacher whose "conceptions and beliefs are consistent with their practice" have had "opportunities to critically reflect on their actions and consider new possibilities for teaching" (p. 2).

Although the effect or practice of teacher reflexivity was not specifically investigated here, comments made by both Karen and Cathy indicated that their teaching role incorporated reflection. Karen, for example, commented that reading and discussing the ideas presented in the hypothetical scenarios caused her to question and think about her practice: "I think there's a fair degree of freedom here in [student] choice…but now you've got me worried…you've got me thinking". Although this highlighted, in part, the effect of the unintended and double hermeneutic consequences of a researcher interacting with a practitioner (see Sect. 4.1.3), it also suggested that when faced with new ideas, Karen would question and re-assess her practices. Cathy also indicated that reflection was an important aspect of her role as a teacher. She described her work environment as one which encouraged an "open minded approach and that willingness to [say] okay, so we've done it this way, why not try it another way? If it's successful, great". She indicated that her pedagogy was the result of consistently "try[ing] new things", and assessing these to find the best approach. Kwo and Intrator (2004) indicated that the manner in which Cathy and Karen utilised their power in the classroom reflected what they termed, "inner power", that is, the "power to learn". They suggested that a teacher with such power could be recognised by a classroom practice "featured by her [sic] inquiry orientation, by which she was open to risk-taking, collaboration with her pupils and mobilising learning resources" (p. 287).

In contrast, neither Lisa nor Elizabeth indicated that reflection was an on-going aspect of their teaching practice. Both teachers reported having observed that their students preferred certain forms of learning experiences, for example, Lisa noted that "we do the hands-on experiments because it really draws them [students] in… they like it, they enjoy it, they want to be part of it, they wanna know what it's about", and Elizabeth indicated that the use of "hands-on [activities] and working with a partner is a really good way of drawing our and pushing out and maximising their [the student's] own understanding of some learning". However, these observations were reported in the form of a justification for current pedagogical routines rather than as an indication of a significant degree of on-going reflection.

However, "teachers' professional frames are both individually and socially derived – shaped by experiences as well as by expectations and values" (Thomas and Pederson 2003, p. 322). Like all aspects of human agency, each teacher's pedagogy reflected a unique personal knowledge and understanding of the world around them: their tacit knowledge (Giddens 1984).

8.5   Tacit Knowledge and Congruence

As indicated by Giddens' notion of tacit knowledge (see Sect. 3.3), the teachers were knowledgeable individuals with well developed understandings of the social and cultural expectations of both teachers and students in educational institutions (Giddens 1976; Stake 2001). Coburn (2004) recognised teachers' tacit knowledge as:

deep-seated assumptions about the nature of teaching and learning that are linked to the broader movements in the [school] environment [and which] guide decision making often in preconscious ways [by] framing the range of appropriate action and guiding what 'makes sense' to teachers" (pp. 234–235).

An individual's tacit knowledge is "developed [and up-dated] over time" (Coburn 2004, p. 235). This suggests that the teachers' tacit knowledge of the process of education, and the manner in which teachers and students interact, began to form during their own schooling: "teachers have themselves spent many years as students in schools, during which time they have developed their own beliefs about teaching, many of which are diametrically opposed to those presented to them during their teacher education" (Korthagen 2001a, p. 81). Such tacit knowledge "constitute[s] a strong framework into which teachers tend to try to 'fit' new approaches and ideas" (Coburn 2004, p. 235). As a result, Lortie (2002) contended that "teachers will often reproduce the strategies they have had as primary, secondary and teacher education students" (Miles and Cutter-Mackenzie 2006, p. 141). In light of this, Coburn (2004) described teachers as acting with "bounded autonomy" (p. 234), that is, that a teacher's "knowledge and beliefs provide a framework for pedagogy, knowledge of students, subject matter and the curriculum, and guides the teachers' actions in practice" (Carrington 2010, p. 2). The notion that tacit knowledge does influence the manner in which the teachers viewed new educational ideas and practices was supported by comments made by both Karen and Cathy. Karen, for example, referred to the recent introduction of new government curriculum guidelines (the Victorian Essential Learning Standards, VELS), stating that "when VELS came in…I thought well thank goodness somebody's written some sense…because this program [at EVNP] had then been in operation…and VELS fitted me…or I fitted VELS". Similarly, Cathy noted that the goals of SSP were congruous with her educational goals to "use our environment and care for the environment, and having something special for the year [Grade] 4's" and that therefore the implementation of SSP "naturally slipped into being" as part of a student "leadership" program.

The teachers' tacit knowledge greatly influenced their decisions about whether or not to implement SSP through a socially-critical pedagogy or a vocational/neo-classical pedagogy. As demonstrated by teachers such as Cathy and Elizabeth, each teacher's chosen pedagogy reflected their judgment of the similarity, or "congruence" (Coburn 2004, p. 227), of their tacit knowledge and the environmental and educational ideologies embedded within the ideals and practices of SSP. This was supported by a study of the responses of Californian teachers to new ideas about reading instruction in which Coburn (2004) found that "the greater the congruence of institutional pressures with the teachers' pre-existing beliefs and practices, the more likely the teachers were to incorporate new approaches and influences into their classroom practice in some manner" (p. 227). Both Karen and Cathy demonstrated that their decisions to implement SSP through a socially-critical pedagogy reflected a high degree of similarity between their personal ideologies, and those represented by the goals and pedagogical requirements of SSP. Both of these teachers held strong views regarding the need for, and validity of, environmental education as a legitimate part of primary school learning. Both Cathy and Karen justified their role as a teacher in terms that were consistent with the learning objectives of a socially-critical pedagogy. In addition, both of these teachers indicated that prior to implementing SSP they had already well-established teaching strategies which incorporated elements of a socially-critical pedagogy. In other words, Cathy and Karen were able to effectively implement SSP through a socially-critical pedagogy by simply incorporating the program into their existing teaching practices.

However, teachers' classroom practices often fail to reflect the rhetoric of new educational ideas which do not correspond with their existing views about the subject and/or their well-established teaching strategies (Olson 1992; Sosniak et al. 1994). For example, Lisa demonstrated significant enthusiasm for the environmental ideals of SSP. She identified the anthropocentric values represented by the goals of SSP, and indicated that these matched many aspects of her own environmental beliefs. However, the socially-critical pedagogy recommended by SSP was not entirely consistent with her educational ideology. Although Lisa recognised that her students could benefit from many aspects of a socially-critical pedagogy, she did not believe that undertaking a socially-critical pedagogy would enable her to fulfil her role as a teacher, particularly in terms of ensuring that her students gained the knowledge outlined in the government curriculum followed by her school. Others support the idea that teachers may not choose to employ a teaching strategy that is only partially consistent with their personal ideologies. For example, a study of the practices of an outdoor environmental educator found that it was the "notion of what it meant to be a proper teacher, rather than barriers related to programme, skills, resources, or his [sic] own beliefs in the value of student-centred pedagogy, that were the main deterrents making it difficult…to teach the way he wanted" (Barrett 2007, p. 213). The manner in which Lisa attempted to implement SSP, as science lessons and a vocational/neo-classical pedagogy, represented her perception of the relationships between her beliefs and the environmental and educational ideologies represented by SSP. Lisa's response matches those found by Coburn (2004) to be typical of many teachers who attempt to implement educational changes that have some degree of congruence with their personal ideologies, by incorporating "the messages by assimilating them into their pre-existing practice, rather than making more substantive adjustments" (p. 227). In other words, Lisa used SSP to introduce environmental education to her students, but chose to do this in a manner that fitted into her existing well-established practices. Thus, the gap between the reality of Lisa's practices and the rhetoric of SSP reflected her decision to implement SSP within science lessons and her existing vocational/neo-classical pedagogy. Although Lisa agreed with the environmental goals of SSP, she found that the pedagogical requirements of the program conflicted with her educational ideology, and therefore, also her beliefs about her role as a teacher.

Like Lisa, Elizabeth agreed that the environmental goals of SSP reflected valid social concerns and addressed the future needs of humanity. However, unlike Lisa, she found it difficult to justify these goals as a legitimate part of primary school education. Other studies have shown that teachers' tacit knowledge may include beliefs about what is legitimate learning. For example, although pre-service teachers in Queensland were found to be competent and prodigious users of digital technologies, and quick to make use of technological changes, "they did not align the changing nature of technology with changes in education" (Donnison 2004, p. 22). Donnison (2004) reported that although these pre-service teachers' "lived experience was one that evidenced technological literacy and competence, their future predictions of themselves as teaching professionals suggested limited technological engagement" and that "technology was predicted to be at the periphery of education", and therefore its inclusion "in the classroom mimics the present situation" (p. 26). In other words, this aspect of a teachers' tacit knowledge may contribute to gaps between their classroom practices and the rhetoric of the educational programs they have been directed to implement.

In addition, and like Lisa, Elizabeth found that the socially-critical pedagogy of SSP was incongruous with her personal educational ideology. Elizabeth believed that this pedagogy could not enable her to fulfil her role as a teacher, particularly in terms of the emotional care of her students, and like Lisa, in terms of ensuring that her students gained the curriculum-specified knowledge deemed appropriate by her school. The manner in which Elizabeth implemented SSP reflected the relationship between the goals of SSP and her environmental and educational ideologies: a routinised waste management system that incorporated minimal environmental learning, and which was meticulously designed and instigated to minimise its impact on Elizabeth's traditional daily lessons. In other words, Elizabeth chose to deliberately separate activities related to the implementation of SSP from what she considered to be her legitimate lessons. Thus, the gap between Elizabeth's practices and those advocated by SSP reflected her decision to not implement SSP, as its educational goals and pedagogical requirements conflicted with her environmental and educational ideologies, and therefore, also her beliefs in her role as a teacher. Other studies indicated that Elizabeth's actions were not an uncommon response to educational change. A study of teachers' responses to changes in Tasmanian educational institutions, for example, revealed "a clash between the administrative ideology of economic rationalism and the professional ideology of care" (Easthope and Easthope 2007, p. 2). This "class of ideologies" caused a great deal of teacher anxiety because "teachers were forced to adapt to changes imposed upon them. The attempt to satisfy the requirements of both ideologies created in some teachers' minds a realisation that it was impossible to maintain the level of teaching they previously enjoyed" (p. 12). The study found that although it seemed as though the required changes were being implemented they were actually merely "adopted in addition to the professional ideology that was already in place" (Easthope and Easthope 2007, p. 10), such that, the reality of the teachers' practices did not match the rhetoric of the program they professed to be implementing.

Thus, the teachers' classroom practices reflected, in part, their tacit knowledge which contributed to their "inclination towards reproducing the status quo" (Donnison 2004, p. 28). However, this effect was not restricted to those teachers whose practices represented rhetoric–reality gaps, but to all of the teachers, irrespective of the manner in which they chose to implement SSP. The tacit knowledge held by Cathy and Karen embraced beliefs and values that were not only consistent with those embedded in the goals of SSP and the recommended socially-critical pedagogy, but which also reflected their existing pedagogical preferences and practices. In contrast, the tacit knowledge of teachers such as Lisa and Elizabeth embraced beliefs and values that could not be supported by the implementation of SSP through a socially-critical pedagogy, but which reflected these teachers' preference for their well-established vocational/neo-classical pedagogy.

8.5.1 The Duality of Structure and Agency: A Note about Praxis

As outlined in Sect. 3.3, each person's tacit knowledge incorporates contextually-specific understandings of social expectation and personal obligation (Giddens 1984; Stones 2005). When implementing SSP, the teachers demonstrated that, in line with Giddens' notion of the duality of structure and agency, each teacher's classroom practices not only reflected their tacit knowledge, but also confirmed that tacit knowledge (Giddens 1984). For example, Elizabeth believed that her students were incapable of making appropriate decisions about their learning; a belief that was continuously supported by her students being prevented from either making decisions, or learning how to make decisions, by Elizabeth's teacher-directed vocational/neo-classical pedagogy. In contrast, Karen, for example, believed that her students could effectively direct many aspects of their learning; a belief that was continuously supported by Karen enabling her students to demonstrate their decision making ability through her use of a student-directed socially-critical pedagogy. This not only demonstrated the duality of structure and agency in defining a teacher's pedagogy, but also the importance of praxis.

Praxis, or the notion that "theory building and critical reflection inform our practice and our action, and our practice and our action inform our theory building and critical reflection" (Wink 2000, p. 59) was an important component of the teachers' practices, irrespective of whether or not those practices represented rhetoric–reality gaps. Each teacher justified their educational practice in relation to their personal educational ideology, and justified their educational ideology in relation to their experience of the effects of their practice. Coburn (2004) reported that "teachers' responses to messages are not static" (p. 235). This was supported by Wink's (2000) notion that "in praxis, the ideas which guide action are just as subject to change as action is" (p. 34). The specific role of praxis in the teachers' practices, or their response to the implementation of SSP, is not fully explored here. However, the role of the duality of structure and agency in defining the teachers' practices suggested that any process designed to alter teachers' educational ideologies would need to provide experiences that assisted teachers to critique and adjust their perceptions and practices (McLaren 1995; Morrow and Torres 2002; Sarason 1990; Wink 2000).

8.6 Educational Rhetoric–Reality Gaps and SSP

Consideration of the rhetoric and the reality of the implementation of SSP by the teachers showed that the structural features of the school work environment did not universally constrain or enable the teachers to implement a socially-critical pedagogy. However, it was evident that each teacher's beliefs about the environment and education influenced: their perception of SSP goals; whether or not they embraced SSP principles in their own lives; and the manner in which they chose to implement SSP in their classrooms (see Chaps. 6 and 7). Thus, in the context of the implementation of SSP, the development of the educational rhetoric–reality gaps reflected issues of teacher agency. This being the case, in order to identify ways in which the development of such educational rhetoric–reality gaps may be reduced, it was essential to identify the influence of the teachers' environmental and educational ideologies on their decisions regarding pedagogy.

Most of the teachers believed that the goals of SSP represented an environmental ideology based on anthropocentric values, and that those values were congruous with their personal environmental ideology. Thus, the educational rhetoric–reality gaps represented by the vocational/neo-classical pedagogy employed by some of the teachers did not necessarily reflect the inability of those teachers to reconcile their environmental ideology with the goals of SSP. Despite this, the teachers referred to the environment in different ways: those who practiced either a vocational/neo-classical or liberal-progressive pedagogy consistently referred to *the* environment; whereas those who practiced a socially-critical pedagogy often spoke of *our* environment, or, in relation to their students, *their* environment. These differences reflected a difference in not just the classroom practices of the teachers, but also their educational ideologies.

All of the teachers held strong beliefs about their educational role, particularly in terms of their responsibility with respect to: student emotional wellbeing; student ability; student choice; learning from others; assessment of student learning; and their power and authority classroom. The effect of each teacher's educational ideology was best illustrated by the manner in which they chose to utilise their power and authority in the classroom, and by what they perceived to be the goal of their teaching strategy. Cathy and Karen, for example, directed their authority in the classroom in a manner that supported a socially-critical pedagogy. They each focused on motivating their students through providing a school environment that was highly responsive to their students' interests. They helped their students to develop their understandings and skills through participation in all aspects of learning activities. In contrast, teachers such as Lisa and Elizabeth directed their authority in the classroom through a vocational/neo-classical pedagogy. They each developed learning opportunities that addressed the specific knowledge-based outcomes of the government curriculum. Although they hoped to identify topics that would interest their students, they did not believe that their students had any role to play in choosing, designing or developing appropriate learning experiences. This was their responsibility and role as a teacher.

Each of these approaches not only reflected the respective teachers' educational ideology, but also their well-established teaching strategies, or routines of practice. Cathy and Karen both found the socially-critical pedagogy of SSP to be congruous with their educational ideology and their existing classroom practices. In contrast, Lisa and Elizabeth considered a socially-critical pedagogy to be in conflict with their educational ideology and therefore also their existing classroom practices. The absence of a gap between a teacher's classroom practices and the rhetoric of the new educational program did not reflect that teacher's ability to successfully respond to educational change, but rather their decision to *not* respond to educational change. A teacher's decision to not respond reflected their inability to reconcile the differences between their personal educational ideology and the educational ideology represented by a socially-critical pedagogy. Most importantly, these teachers sought to minimise the rhetoric–reality gap between their personal educational ideology and their chosen pedagogy. In all cases, a teacher's educational ideology was congruous with their chosen pedagogy, even when that pedagogy conflicted with the pedagogical requirements of SSP. All of the teachers chose to implement the ideals of SSP in a manner consistent with their educational ideology, that is, their tacit knowledge of what it meant to be teacher (see also Coburn 2004; Korthagen 2001a, b; Lortie 2002).

8.7   Ontological Security and Educational Change

Ontological security was considered by Giddens to be an individual's unconscious safety system—the desire to avoid negative emotions such as anxiety or guilt (see Sect. 3.9). In other words, people act with some reference to feelings, and in accordance with beliefs, values and attitudes. This implies that people temper their actions with some reference to feelings, and assert their agency in accordance with their most strongly held beliefs, values and attitudes (Stones 2005). A detailed investigation of each teacher's perception of what constituted ontological security, and the relationship between that teacher's agency and ontological security, was not undertaken. However, in light of the apparent strong link between each teacher's educational ideology, belief in their role as a teacher, and their choice of pedagogy, it is evident that each teacher employed their agency to undertake their teaching role in a manner that supported their beliefs, and in so doing, enabled them to maintain some sense of ontological security. Thus, Giddens' notion of ontological security has significant implications for finding ways in which to reduce the prevalence of educational rhetoric–reality gaps. Understanding and effectively addressing issues of teacher ontological security is an essential component of any policy or program that aims to guide educational change.

References

Amérigo, M., Aragonés, J. I., de Fructos, B., Sevillano, V., & Cortés, B. (2007). Underlying dimensions of ecocentric and anthropocentric environmental beliefs. *The Spanish Journal of Psychology, 10*(1), 97–103.

Angus, M., Olney, H., & Ainley, J. (2007). *In the balance. The future of Australia's primary schools*. Australian Primary Principal's Association. www.appa.asn.au/reports/In-the-balance. pdf. Accessed 1 Oct 2014.

Bailey, B. (2000). The impact of mandated change on teachers. In N. Bascia & A. Hargreaves (Eds.), *The sharp edge of educational change* (pp. 112–129). New York: Routledge.

Barrett, M. J. (2007). Homework and fieldwork: Investigations into the rhetoric-reality gap in environmental education research and pedagogy. *Environmental Education Research, 13*(2), 209–223.

Bishop, A. J. (2008). Values in mathematics and science education: Similarities and differences. *The Montana Mathematics Enthusiast, 5*(1), 47–58.

Bishop, A. J., Clarke, B., Corrigan, D., & Gunstone, D. (2005). Teachers' preferences and practices regarding values in teaching mathematics and science. In P. Clarkson, A. Downton, D. Gronn, M. Horne, A. McDonough, R. Pierce, et al. (Eds.), *Building connections: Theory, research and practice. Proceedings of mathematics education research group of Australasia* (pp. 153–160). Melbourne: RMIT University.

Bishop, A. J., Gunstone, D., Clarke, B., & Corrigan, D. (2006). Values in mathematics and science education: Researchers' and teachers' views on the similarities and differences. *For the Learning of Mathematics, 26*(1), 7–11.

Brandt, R. S. (2003). How new knowledge about the brain applies to social and emotional learning. In M. J. Elias, H. Arnold, & C. Steiger-Hussey (Eds.), *EQ + IQ = Best leadership practices for caring and successful schools* (pp. 57–70). Thousand Oaks: Corwin Press.

Brunker, N. (2007, November 25–29). *Primary schooling and children's social emotional wellbeing: A teacher's perspective. In Conference of the Australian Association for Research in Education. Research impacts: Proving or improving? University of Notre Dame, Fremantle, Australia* (pp. 1–11).

Butcher, J., & McDonald, L. (2007). *Making a difference: Challenges for teachers, teaching, and teacher education*. Rotterdam: Sense Publishers.

Caprara, G. V., Barbaranelli, C., Pastorelli, C., Bandura, A., & Zimbardo, P. G. (2000). Prosocial foundations of children's academic achievement. *Psychological Science, 11*(4), 302–306.

Carrington, S. (2010). Cultivating teachers' beliefs, knowledge and skills for leading change in schools. *Australian Journal of Teacher Education, 35*(1), 1–13.

Chin, C., Leu, Y.-C., & Lin, F.-L. (2001). Pedagogical values, mathematics teaching, and teacher education: case studies of two experienced teachers. In F.-L. Lin & T. J. Cooney (Eds.), *Making sense of mathematics teacher education*. Dordrecht: Kluwer Academic Publishers.

Clarkson, P., Downton, A., Gronn, D., Horne, M., McDonough, A., Pierce, R., et al. (2005, July 7–9). Building connections: Theory, research and practice. In *Proceedings of the annual conference of the Mathematics Education Research Group of Australasia* (pp. 153–160). Melbourne: RMIT University.

Coburn, C. E. (2004). Beyond decoupling: Rethinking the relationship between the institutional environment and the classroom. *Sociology of Education, 77*, 211–244.

Cotton, D. R. E. (2006). Implementing curriculum guidance on environmental education: The importance of teachers' beliefs. *Journal of Curriculum Studies, 38*(1), 67–83.

Cutter, A. (1998). *Integrated pre-service environmental education: The abandonment of knowledge*. Honours, Griffith University, Australia.

Devine, D. (2000). Constructions of childhood in school: Power, policy and practice in Irish education. *International Studies in Sociology of Education, 10*(1), 23–41.

Donnison, S. (2004). The 'digital generation', technology, and educational change: An uncommon vision. In B. Bartlett, F. Bryer, & D. Roebuck (Eds.), *Educating: Weaving research into prac-*

tice (Vol. 2, pp. 22–31). Brisbane: School of Cognition, Language and Special Education, Griffith University.

Dunlop, R. E., van Liere, K. D., Merting, A. G., & Jones, R. E. (2000). Measuring endorsement of the new environmental paradigm: A revised NEP scale. *Journal of Social Issues, 5,* 425–442.

Easthope, C., & Easthope, G. (2007). Teachers' stories of change: Stress, care and economic rationality. *Australian Journal of Teacher Education, 32*(1), 1–16.

Evans, B. (1987). Agency and structure: Influences on a principal's initiation of change in school practices. *Australian Journal of Education, 31*(3), 272–283.

Evans, L. (2000). The effects of educational change on morale, job satisfaction and motivation. *Journal of Educational Change, 1*(173-192).

Fien, J. (1993). *Education for the environment: Critical curriculum theorising and environmental education.* Geelong: Deakin University Press.

Fien, J. (1999). Towards a map of commitment: A socially critical approach to geographical education. *International Research in Geographical and Environmental Education, 8*(2), 140–158.

Fook, L., Repetti, R. L., & Ullman, J. B. (2005). Classroom social experiences as predictors of academic performance. *Developmental Psychology, 41*(2), 319–327.

Freire, P. (1972). *Pedagogy of the oppressed.* Harmondsworth: Penguin.

Giddens, A. (1976). *New rules of sociological method.* New York: Hutchinson.

Giddens, A. (1979). *Central problems in social theory: Action, structure and contradiction in social analysis.* London: Macmillan.

Giddens, A. (1984). *The constitution of society.* Cambridge: Polity Press.

Giroux, H. A. (1988). *Teachers as intellectuals.* New York: Bergin and Garvey.

Goodwin, B. (2007). *Using political ideas* (5th ed.). London: John Wiley.

Gough, A. (2007, November 25–29). *Beyond convergence: Reconstructing science/environmental education for mutual benefit.* In Annual conference of the Australian Association for Research in Education (AARE), Freemantle, Australia.

Gray, J. (2009). *False dawn: The delusions of global capitalism.* London: Granta.

Greenall Gough, A., & Robottom, I. (1993). Towards a socially critical environmental education: Water quality studies in a coastal school. *Journal of Curriculum Studies, 25*(4), 301–316.

Grun, M. (1996). An analysis of the discursive production of environmental education: Terrorism, archaism and transcendalism. *Curriculum Studies, 43*(3), 329–339.

Hargreaves, A. (1998). The emotional practice of teaching. *Teaching and Teacher Education, 14*(8), 835–854.

Harvey, D. (2005). *A brief history of neoliberalism.* Oxford: Oxford University Press.

Jickling, B. (2005). The wolf must not be made a fool of. In P. Tripp & L. Muzzin (Eds.), *Teaching as activism: Equity meets environmentalism* (pp. 35–46). Montréal: McGill-Queen's University Press.

Kearney, P., Plax, T. G., Richmond, V. P., & McCroskey, J. C. (1984). Power in the classroom IV: Alternatives to discipline. *Communication Yearbook, 8,* 724–746.

Kearney, P., Plax, T. G., Richmond, V. P., & McCroskey, J. C. (1985). Power in the classroom III: Teacher communication techniques and messages. *Communication Education, 34*(1), 19–28.

Keller, D. R. (2010). *Environmental ethics: The big questions.* Chichester: Wiley-Blackwell.

Kelly, T. M. (1986). Discussing controversial issues: Four perspectives on the teacher's role. *Theory and Research in Social Education, 14*(2), 309–327.

Kemmis, S., Cole, P., & Suggett, D. (1983). *Orientations to curriculum and transition: Towards the socially-critical school.* Melbourne: Victorian Institute of Secondary Education.

Korthagen, F. (2001a). In search of the essence of a good teacher: Towards a more holistic approach in teacher education. *Teaching and Teacher Education, 20,* 77–97.

Korthagen, F. (2001b). *Linking practice with theory: The pedagogy of realistic teacher education.* Mahwah: Lawrence Erlbaum.

Kwo, O. W. Y., & Intrator, S. M. (2004). Uncovering the inner power of teachers' lives: Towards a learning profession. *International Journal of Educational Research, 41,* 281–291.

Lee, J. C.-K. (1993). Geography teaching in England and Hong Kong: Contributions towards environmental education. *International Research in Geographical and Environmental Education, 2*(1), 25–40.

Lortie, D. (2002). *Schoolteacher: A sociological study* (2nd ed.). Chicago: University of Chicago.

Lucas, A. M. (1980). Science and environmental education: Pious hopes, self praise and disciplinary chauvinism. *Studies in Science Education, 7*, 1–21.

Malecki, C. K., & Elliott, S. N. (2002). Children's social behaviours as predictors of academic achievement: A longitudinal analysis. *School of Psychological Quarterly, 17*(1), 1–23.

Manno, J. (2004). Political ideology and conflicting environmental paradigms. *Global Environmental Politics, 4*(3), 155–159.

Matthews, A. (2005). Mainstreaming transformative teaching. In P. Tripp & L. Muzzin (Eds.), *Teaching as activism: Equity meets environmentalism* (pp. 95–105). Montréal: McGill-Queen's University Press.

McCroskey, J. C., Richmond, V. P., Plax, T. G., & Kearney, P. (1985). Power in the classroom V: Behaviour alteration techniques, communication, training and learning. *Communication Education, 34*(3), 214–226.

McLaren, P. (1995). *Critical pedagogy and predatory culture* (Oppositional politics in a postmodern era). New York: Routledge.

Michalak, M. (2007). In search of the conditions of teachers' success. In J. Butcher & L. McDonald (Eds.), *Making a difference: Challenges for teachers, teaching, and teacher education* (pp. 69–82). Rotterdam: Sense Publishers.

Miles, R., & Cutter-Mackenzie, A. (2006). Environmental education: Is it really a priority in teacher education? In S. Wooltorton & D. Marinova (Eds.), *Sharing the wisdom for our future: Environmental education in action* (pp. 1–6). Sydney: Australian Association for Environmental Education.

Milfont, T. L., & Duckitt, J. (2004). The structure of environmental attitudes: A first- and second-order confirmatory factor analysis. *Journal of Environmental Psychology, 24*, 289–303.

Mogensen, F. (1997). Critical thinking: A central element in developing action competence in health and environmental education. *Health Education Research, 12*(4), 429–436.

Morrow, R. A., & Torres, C. A. (2002). *Reading Freire and Habermas. Critical pedagogy and transformative social change.* New York: Teachers College Press.

Olson, J. M. (1992). *Understanding teaching: Beyond expertise.* Buckingham: Open University Press.

Phillips Mankes, M. (1997). *Classroom power relations. Understanding student-teacher interaction.* Mahwah: Lawrence Erlbaum.

Rai, J. S., Thorheim, C., Dorjderem, A., & Macer, D. (2010). *Universalism and ethical values for the environment* (Report of the ethics and climate change in Asia and the Pacific Project Working Group). United Nations Educational, Scientific and Cultural Organisation (UNESCO). http://unesdoc.unesco.org/images/0018/001886/188607e.pdf. Accessed 1 Oct 2014.

Richmond, V. P. (1990). Communication in the classroom. *Communication Education, 39*, 181–195.

Richmond, V. P., & McCroskey, J. C. (1984). Power in the classroom: Power and learning. *Communication Education, 33*, 125–136.

Richmond, V. P., & McCroskey, J. C. (1992). *Power in the classroom. Communication, control and concern.* Mahwah: Lawrence Erlbaum.

Robertson, C. L., & Krugly-Smolska, E. (1997). Gaps between advocated practices and teaching realities in environmental education. *Environmental Education Research, 3*(3), 311–326.

Sarason, S. B. (1990). *The predictable failure of educational reform: Can we change course before it's too late?* San Francisco: Jossey-Bass.

Schugurensky, D. (2002). Transformative learning and transformative politics. The pedagogical dimension of participatory democracy and social action. In E. O'Sullivan, A. Morrell, & M. A. O'Connor (Eds.), *Expanding the boundaries of transformative learning. Essays on theory and praxis* (pp. 59–76). New York: Palgrave.

Schultz, P. W. (2001). The structure of environmental concern: Concern for self, other people, and the biosphere. *Journal of Environmental Psychology, 21*, 327–339.

Scott, W., & Gough, S. (2004). *Key issues in sustainable development and learning: A critical review*. London: Routledge.

Scott, W., & Oulton, C. (1999). Environmental education: Arguing the case for multiple approaches. *Educational Studies, 25*(1), 89–97.

Short, K., & Burke, C. (1996). Examining our beliefs and practices through inquiry. *Language Arts, 73*(2), 97–104.

Simmons, D. A. (1991). Are we meeting the goal of responsible environmental behaviour? An examination of nature and environmental education center goals. *Journal of Environmental Education, 22*(3), 16–21.

Sosniak, L. A., Ethington, C. A., & Varelas, M. (1994). The myth of progressive and traditional orientations: Teaching mathematics without a coherent point of view. In I. Westbury, C. A. Ethington, L. A. Sosniak, & D. P. Baker (Eds.), *In search of more effective mathematics education: Examining data from the IEA second international mathematics study* (pp. 95–112). Norwood: Ablex.

Spring, J. (2004). *How educational ideologies are shaping global society: Intergovernmental organisations, NGOs, and the decline of the nation-state*. Mahwah: Lawrence Erlbaum.

Stake, R. E. (2001). The case study method in social inquiry. In N. K. Denzin & Y. S. Lincoln (Eds.), *The American tradition in qualitative research* (Vol. II, pp. 131–138). London: Sage.

Sterling, S. (2003). *Whole systems thinking as a basis for paradigm change in education: Explorations in the context of sustainability*. PhD, University of Bath, UK.

Stones, R. (Ed.). (2005). *Structuration theory* (Traditions in social theory). London: Palgrave MacMillan.

Sunderlin, W. D. (2003). *Ideology, social theory and the environment*. Lanham: Rowman and Littlefield.

Swain, A. (2005). *Education as social action*. London: Palgrave MacMillan.

Thomas, J., & Pederson, J. (2003). Reforming elementary science teacher preparation: What about extant teaching beliefs? *Journal of School Science and Mathematics Association, 103*(7), 319–321.

Thompson, S. C. G., & Barton, M. (1994). Ecocentric and anthropocentric attitudes towards the environment. *Journal of Environmental Psychology, 14*, 149–157.

Trautmann, N. M., & MacKinster, J. G. (2005, January 19–23). *Teacher/scientist partnerships as professional development: Understanding how collaboration can lead to inquiry*. Paper presented at the International Conference of the Association for the Education of Teachers of Science, Colorado Springs, USA. http://ei.cornell.edu/pubs/AETS_CSIP_%2005.pdf. Accessed 1 Oct 2014.

Turner, J. H. (2003). Structuration theory: Anthony Giddens. In *The structure of sociological theory* (7th ed., pp. 476–490). Belmont: Wadsworth.

Vander Staay, S. L., Faxon, B. A., Meischen, J. E., Kolesnikov, K. T., & Ruppel, A. D. (2009). Close to the heart: Teacher authority in a classroom community. *College Composition and Communication, 61*(2), 262–282.

Vilkka, L. (1997). *The intrinsic value of nature* (Value inquiry book series, Vol. 59). Amsterdam: Editions Rodopi.

Vincent, A. (2010). *Modern political ideologies*. Chichester: Wiley-Blackwell.

Walker, K. (1997). Challenging critical theory in environmental education. *Environmental Education Research, 3*(2), 155–162.

Wink, J. (2000). *Critical pedagogy: Notes from the real world* (2nd ed.). New York: Addison Wesley Longman.

Chapter 9
The Dance of Structure and Agency—Socially-Critical Environmental Education in Primary Classrooms

The understanding that it is not possible to sustain current human–environment relationships, due to the potential catastrophic social and environmental consequences of the unmitigated use of natural resources combined with exponential population growth, has led to global calls to transform the way human societies operate. Any journey of social transformation begins with the willingness and ability to question the philosophy upon which current cultural practices are founded. This requires the institutions that define and are defined by the predominant cultural values of a society, and which actively support the continuance of those cultural values, to question their role in society and to find ways in which to empower communities and individuals to re-assess the ways in which they act upon those values. The Decade of Education *for* Sustainable Development (2005–2014) prompted educational institutions to question how they contribute to the continuance of well-established unsustainable human–environment relationships, and to find ways in which to empower educators to embrace the practices that would empower students to actively participate in social transformation towards sustainable living. However, the development of educational rhetoric–reality gaps during the implementation of programs that provide opportunities to question the role of education and facilitate new practices suggest that educational transformation is problematic, and that more-effective change processes are required. In order to better inform the process of educational transformation, it is essential to better understand the process of education. Research that is informed by the holistic view of social interaction provided by Anthony Giddens' theory of structuration has the potential to reveal new understandings of teachers' pedagogical practices, and to identify possible intervention points for assisting teachers to change these practices in order to reduce the development of educational rhetoric–reality gaps in the future.

© Springer International Publishing Switzerland 2016
J. Edwards, *Socially-critical Environmental Education in Primary Classrooms*,
International Explorations in Outdoor and Environmental Education 1,
DOI 10.1007/978-3-319-02147-8_9

9.1   Questioning Current Practices

The establishment of the Decade of Education *for* Sustainable Development (DESD; 2005–2014) incorporated recommendations from the United Nations Educational, Scientific and Cultural Organisation (UNESCO) that emerged from the global community's questioning of the potential of the well-established cultural values that regulate human–environment relationships to affect the Earth's ability to support the goals that such cultural values strive to achieve. The DESD was founded on the understanding that the ability of the Earth's global community to thrive into the future will be determined, in part, by the way in which today's educators prepare students for their future decision-making roles. In light of this, the goals of DESD demanded urgent and radical changes to be made to education in order to better prepare humanity for the potentially detrimental and planetary-scale effects of the human–environment relationships that sustain the industrialised world. In other words, the goals of DESD demanded educational change in order to facilitate social transformation. Enacting such socially-transformative education required educational institutions to question their role in society, and to assess whether or not their well-established practices unquestionably supported the continuance of the predominant cultural values, or effectively prepared students for their future in a society shaped by ever increasing rates of change. This also required educators to re-evaluate their role, not just in relation to the practical aspects of classroom teaching, but also in terms of how those practices defined the purposes of education as shaped by society, and through which society has been, is, and will in the future be shaped.

In Australia, the educational change advocated by DESD was encouraged through the development of the Sustainable Schools Program (SSP), an educational framework that embraced the ideals of Education *for* Sustainable Development (ESD). The effective implementation of SSP required educational institutions to significantly alter their organisational structure in order to facilitate opportunities for both teachers and students to work and learn within the context of a sustainable environment (see Chap. 2). However, the socially-transformative education advocated by DESD (and SSP) also required educators to question their practices and the role they play in shaping a society. The school principals were adamant that their decision to implement SSP reflected the single most important and urgent aspect of their educational leadership role; to bring about pedagogical reform, most particularly, to reduce the prevalence of the out-dated but entrenched vocational/neo-classical instruction in their schools (see Sect. 6.2). Thus, the success of SSP to guide the development of socially-transformative education depended most on the ability and willingness of teachers to embrace new pedagogical practices, in particular, a socially-critical pedagogy (see Chap. 2).

The classroom practices of several teachers in Victorian primary schools (the teachers' stories given in Chap. 5) indicated that the implementation of SSP was accompanied by the development of educational rhetoric–reality gaps; differences between the reality of the vocational/neo-classical and liberal-progressive classroom pedagogies practiced by some of the teachers and the rhetoric of SSP and the socially-critical pedagogy required to achieve the intended future-oriented and socially transformative goals of ESD. Irrespective of classroom experience or the degree of professional and collegial support, the majority of the teachers participating in the process of educational change framed by SSP were most likely to implement a vocational/neo-classical pedagogy; an approach known to be unsupportive of the socially-transformative educational goals of SSP. In other words, the reality of most of the teachers' classroom practices did not reflect the rhetoric of SSP, and therefore represented educational rhetoric–reality gaps.

The teachers' stories attest to the dynamics and complexities of the role of a teacher in a primary classroom. These stories reveal the plurality of the teachers' ideologies and experiences, and hint at the wide range of possibilities and situations within the context of an educational institution that could contribute to the development of educational rhetoric–reality gaps (Flyvbjerg 2004; Smith 1995). Above all else, the teachers' stories contain the message that there is unlikely to be a single, or simple, reason for the development of rhetoric–reality gaps in response to the implementation of any program that strives for educational change. Thus, in order to reduce the prevalence of educational rhetoric–reality gaps during the implementation of future programs it is essential to first understand the specific characteristics of such gaps. This requires questioning how an understanding of such an issue is best developed in order to effectively inform the process of transformation, and requires finding ways in which to investigate the complex social environment of the educational institutions in which teachers work.

9.2 Investigating Educational Practices: Giddens' Theory of Structuration

Traditional approaches to educational research have viewed teachers as either the primary determinants of their actions in the classroom, or as subjects whose actions are mostly directed by social structural forces beyond their control. This has, in effect, compartmentalised educational research findings into two groups, namely, those that address subjective factors and those that address objective factors. Giddens' theory of structuration (as outlined in Chap. 3) aims to more holistically describe the ontology of social processes, such as the practices of education. According to Giddens', social interaction reflects the interrelatedness of ontological elements that encompass aspects of socio-cultural structures and human agency (or forms of human knowledgeability; Fig. 3.4). Furthermore, Giddens' notion of the duality of structure and agency suggests that these ontological elements both contribute to, and are influenced by, each other element as they are "instantiated in social processes" (Giddens 1984, p. 25). A research process that is guided by a framework that embraces Giddens' notion of the duality of structure and agency provides opportunities to develop an understanding, of any social issue, of greater breadth and depth than research guided by a framework focused solely on either socio-cultural structures or human agency. In light of this, the use of Giddens' theory of structuration, an approach yet to be established within the field of educational research, offers the potential to reveal new insights into the issues faced by teachers when asked to change their well-established educational practices in order to implement new programs, such as SSP and the associated socially-critical pedagogy.

Effectively employing structuration requires researchers to contribute to refining and validating the applicability of Giddens' ideas to the field of education, and to develop ontological and epistemological frameworks that provide research pathways that facilitate opportunities to improve insights and understandings of educational processes. Developing such frameworks and pathways presents many opportunities for future innovative research. Giddens (1989) noted that structuration is not intended to be imported "*en bloc*" into any single empirical research and emphasised "that the theory should be utilised only in a selective way in empirical work and should be seen more as a sensitizing device than as providing detailed guidelines for research procedure" (p. 294, original italics). It is impracticable, and most likely impossible, to comprehensively investigate every single aspect of a specific human action or social interaction. Thus, the design of any research founded on the ideals of structuration must reflect careful consideration of: which of the structuration ontological elements contribute most to a specific research issue; and how to most effectively investigate these ontological elements (see Chap. 4).

There is little doubt that Giddens' notion of a duality of structure and agency presents researchers with methodological dilemmas, not the least of which is the question of whether or not it is possible to effectively identify and interpret any aspect of either structure or agency given that it was instantiated only in the moment of action. Although the answer to this is undoubtedly no, this does not diminish the value of attempting to understand the elements of both structure and agency that contribute to an action, and most importantly, the relationships between these. This requires careful consideration of how to effectively reveal the ways in which the elements of structure and agency both shape, and are shaped by, the social interactions that define a specific research issue.

Giddens' theory of structuration can effectively inform a research framework that highlights the critical hermeneutic and structural ontological elements that interrelate to define educational practices. Understanding these interrelationships is key to identifying intervention points for preventing the development of educational rhetoric–reality gaps when implementing change. Thus, much can be learned about educational practices through developing an understanding of interrelationships between elements of both structure and agency, and in particular, how such relationships are reflected by the teaching and learning of different outcomes, in different classrooms, and by different teachers.

9.3   SSP and Rhetoric–Reality Gaps

The implementation of SSP as a program that facilitates socially-transformative ESD was accompanied by the development of educational rhetoric–reality gaps; differences between the reality of the vocational/neo-classical and liberal-progressive classroom pedagogies practiced by some of the teachers and the rhetoric of SSP and the advocated socially-critical pedagogy. The research that is discussed in this book was informed by a structuration ontology-in-situ framework (see Figs. 4.1 and 4.2), and aimed to investigate the most critical elements of socio-cultural structures and aspects of human knowledgeability that prevented the effective implementation of SSP.

9.3.1 Socio-cultural Structures

Many of the teachers who were required to implement SSP attributed their 'inability' to employ a socially-critical pedagogy to one or more of the ontological structural elements common to any educational institutional environment, including for example: the lack of appropriate learning spaces; the inflexibility of certain school routines; a lack of time; or the scarcity of essential teaching and learning equipment. However, as predicted by Giddens' notion of the duality of structure and agency, although a teacher's practices could certainly be influenced by the structural elements of the school environment, a teacher's practices could also significantly influence such structural elements. The teachers who employed a socially-critical pedagogy demonstrated that the potential 'constraints' of such structural elements, as reported by teachers who employed a vocational/neo-classical or liberal-progressive pedagogy, were substantially reduced, or even eliminated, by their use and appropriate adaptation of a socially-critical pedagogy. Thus, the structural ontological elements of the teachers' work environments did not universally constrain, or enable, the teachers' ability to implement a socially-critical pedagogy. As such, neither the presence, nor the absence, of rhetoric–reality gaps in the implementation of SSP could be predicted by a teachers' perceptions of the structural ontological elements of a school work environment (see Chap. 7). This indicated that the teachers' practices, including the vocational/neo-classical, liberal-progressive and socially-critical pedagogies, were most influenced by ontological elements related to aspects of the teachers' agency.

9.3.2 Human Knowledgeability

In order to develop an understanding of the role of human agency in the development of educational rhetoric–reality gaps in the implementation of SSP it was essential to investigate the ontological elements of the teachers' knowledgeability, that is, their unconscious, conscious and non-conscious forms of knowledge (see Sect. 3.3). However, identifying these forms of knowledge, and most importantly, the disparities between them, was somewhat problematic (see Sect. 4.4.1). The use of hypothetical scenarios in interviews proved to be the key to identifying the subtleties of the teachers' knowledgeability that were most relevant to the development of the rhetoric–reality gaps in the implementation of SSP.

The teachers interpreted the future-oriented and socially-transformative goals of SSP to support anthropocentric environmental values that were, most importantly, congruous with their personal environmental beliefs. This suggested that the teachers did not consider the goals of SSP to be an ideological barrier to implementing the program. The teachers also held strong beliefs about their role as educators, particularly in terms of their responsibility regarding: student emotional wellbeing; student ability; student choice; learning from others; assessment of student learning; and the employment of their power and authority in the classroom. Each teacher embraced a pedagogy that was congruous with their beliefs regarding these responsibilities.

In the context of the implementation of SSP, education ideological rhetoric–reality gaps reflected a conflict between the teachers' beliefs regarding the practices that defined their role as a teacher, and the educational ideological values embedded in the practice of a socially-critical pedagogy. Irrespective of the structural ontological elements of their work environment, each teacher used their agency in a manner that enabled them to maintain their chosen, and in these cases, their preferred, classroom pedagogy. Although such rhetoric–reality gaps developed in response to the introduction of a program that aimed to facilitate educational change, these gaps were not, strictly speaking, caused by the teachers' attempts to *implement* that change. The teachers who demonstrated classroom practices congruous with the socially-critical pedagogy advocated by SSP had not significantly altered their practices in response to SSP, because they were simply continuing to enact their ideologically preferred and well-established practices. Similarly, the vocational/neo-classical and liberal progressive pedagogical practices that defined the rhetoric–reality gaps in the implementation of SSP did not reflect the teachers' attempts to *implement* educational change, but instead, reflected differences between those teachers' ideologically preferred and well-established practices and those advocated by SSP. In all cases, the teachers used their agency to continue to enact their ideologically preferred and well-established practices. In all cases, the teachers accommodated any structural component of their work environment in order to implement the requirements of SSP only if they believed that such action also supported their educational ideology (see Chap. 8).

For many of the teachers, the implementation of the socially-critical pedagogy advocated by SSP would have contradicted their notion of what it meant to be a teacher, and in turn, reduced their ability to maintain a sense of ontological security in their work environment. Thus, both the presence and the absence of educational rhetoric–reality gaps in the implementation of SSP could be attributed to the actions that the teachers took towards fulfilling what they believed to be their role as a teacher, their personal educational ideology, and therefore, the actions that the teachers took towards maintaining their sense of ontological security at work. Each teacher's educational ideology, or tacit knowledge of what it meant to be a teacher, not only predicted their classroom practices, but those practices also re-confirmed their tacit knowledge. This duality contributed to the propensity of the teachers to implement SSP through their existing well-established practices, and supported the notion that "the beliefs teachers hold with regard to learning and teaching determine their actions" (Korthagen 2001, p. 81). In light of this, the rhetoric–reality gaps that accompanied the implementation of SSP, and as defined by the teachers' pedagogical practices, can be more accurately defined as 'education ideological rhetoric–reality gaps.'

9.4 Reducing Education Ideological Rhetoric–Reality Gaps

Figure 9.1 shows that, within the socio-cultural structures of a primary school environment, and in the context of the implementation of SSP, the teachers' practices were driven by their educational ideology. Of the ontological elements critical to the development of these gaps, some formed relationships best described as a duality, including: routines of practice and educational ideology; routines of practice and ontological security; educational ideology and unconscious motives; and educational ideology and ontological security.

In order to reduce the development of the education ideological rhetoric–reality gaps defined by the relationships given in Fig. 9.1, educational policies and school programs that aim to initiate and/or facilitate educational change must address issues of educational ideology and teacher ontological security. Most importantly, this requires programs for educational change, particularly pedagogical change, to incorporate opportunities for teachers to identify ways in which to relate new pedagogical practices to their existing personal educational ideology, that is, to relate new pedagogies to their belief in what it means to be a teacher. The provision of opportunities to successfully address this issue is the key to reducing the prevalence of educational ideological rhetoric–reality gaps in the implementation of programs to facilitate educational change, particularly in relation to programs that advocate pedagogical change. The development of education programs that more effectively facilitate pedagogical change would benefit from research with a focus on identifying ways in which teachers' educational ideology both shapes, and is shaped by, their classroom practices. This would assist to identify ways in which to intervene in this ideology–pedagogy duality, in order to facilitate pedagogical change that is both supported by, and supportive of, teachers' educational ideology and teachers' feelings of ontological security.

9.4.1 The Way Forward

Human–environment relationships both define, and are defined by, the predominant cultural values of a society. If human–environment relationships are to be responsive and adaptive to rapidly changing social and environmental conditions into the future, educational institutions and educational processes must empower students to actively participate in a responsive and adaptive society. This challenge requires educational institutions and individual educators to establish processes that embrace

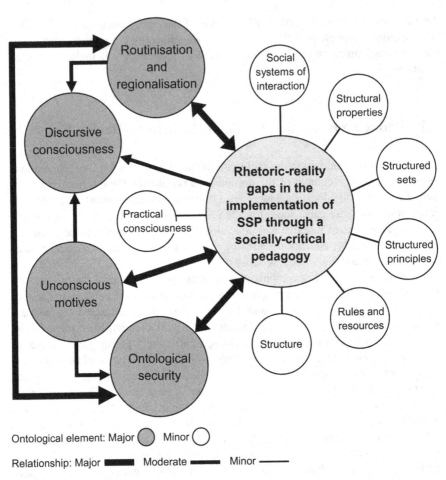

Fig. 9.1 The duality of structure and agency in the development of education ideological rhetoric–reality gaps in the implementation of SSP

change as a professional normality. The development of the education ideological rhetoric–reality gaps in the implementation of SSP indicated that achieving educational change requires a great deal of understanding of educational practice. Just as educational institutions and individual educators must learn to be adaptive and embrace change, so too must the field of educational research and individual researchers. As the role of education in society is questioned and challenged, the forms of understanding that have traditionally informed educational policy and practice must also be questioned and challenged. Educational researchers must also be responsive and adaptive. New ideas and approaches to educational research, such as those informed by the ideals of Giddens' theory of structuration and the duality of structure and agency, are essential for providing the new perspectives and understandings that may lead to identifying ways in which to more effectively facilitate lasting educational change.

References

Flyvbjerg, B. (2004). Five misunderstandings about case-study research. In C. Seale, G. Gobo, J. F. Gubrium, & D. Silverman (Eds.), *Qualitative research practice* (pp. 420–434). Thousand Oaks: Sage.

Giddens, A. (1984). *The constitution of society*. Cambridge: Polity Press.

Giddens, A. (1989). A reply to my critics. In D. Held & J. Thompson (Eds.), *Social theory of modern societies: Anthony Giddens and his critics* (pp. 249–301). Wiltshire: Redwood Burn.

Korthagen, F. (2001). In search of the essence of a good teacher: Towards a more holistic approach in teacher education. *Teaching and Teacher Education, 20*, 77–97.

Smith, A. (1995, October 12–13). Casing the joint: Case study methodology in organisational research. In *Vocational education and training research conference, research and practice: Building the links, Edmund Barton Centre, Melbourne* (pp. 5–8). Melbourne: OTFE.

Index

© Springer International Publishing Switzerland 2016
J. Edwards, *Socially-critical Environmental Education in Primary Classrooms*,
International Explorations in Outdoor and Environmental Education 1,
DOI 10.1007/978-3-319-02147-8

Printed in the United States
By Bookmasters